T0075050

Cambridge Elements

Elements in Decision Theory and Philosophy
edited by
Martin Peterson
Texas A&M University

EVIDENTIAL DECISION THEORY

Arif Ahmed
University of Cambridge

CAMBRIDGE
UNIVERSITY PRESS

University Printing House, Cambridge CB2 8BS, United Kingdom

One Liberty Plaza, 20th Floor, New York, NY 10006, USA

477 Williamstown Road, Port Melbourne, VIC 3207, Australia

314–321, 3rd Floor, Plot 3, Splendor Forum, Jasola District Centre, New Delhi – 110025, India

103 Penang Road, #05–06/07, Visioncrest Commercial, Singapore 238467

Cambridge University Press is part of the University of Cambridge.

It furthers the University's mission by disseminating knowledge in the pursuit of education, learning, and research at the highest international levels of excellence.

www.cambridge.org
Information on this title: www.cambridge.org/9781108713399
DOI: 10.1017/9781108581462

First published 2021

A catalogue record for this publication is available from the British Library.

ISBN 978-1-108-71339-9 Paperback
ISSN 2517-4827 (online)
ISSN 2517-4819 (print)

Evidential Decision Theory

Elements in Decision Theory and Philosophy

DOI: 10.1017/9781108581462
First published online: September 2021

Arif Ahmed
University of Cambridge

Author for correspondence: Arif Ahmed, ama24@hermes.cam.ac.uk

Abstract: Evidential Decision Theory is a radical theory of rational decision-making. It recommends that instead of thinking about what your decisions *cause*, you should think about what they *reveal*. This Element explains in simple terms why thinking in this way makes a big difference, and argues that doing so makes for *better* decisions. An appendix gives an intuitive explanation of the measure-theoretic foundations of Evidential Decision Theory.

Keywords: decision theory, rational choice, subjective probability, Newcomb's Problem, randomization

ISBNs: 9781108713399 (PB), 9781108581462 (OC)
ISSNs: 2517-4827 (online), 2517-4819 (print)

Contents

1 Introduction

1.1 Normative Decision Theory: What It Is

People make dozens, maybe hundreds of decisions per day. In view of all this practice it is alarming how haphazardly it goes. Most people would drive 15 minutes to save $5 on a $15 jacket but not to save $5 on a $125 calculator.[1] A recent US President based vital decisions on the advice of an astrologer.[2] The Naskapi of Labrador decided where to hunt by the cracks and spots that appeared when they held caribou bones over fire.[3] Rome was supposedly founded on the Palatine Hill because of how many birds Romulus could see from it.[4]

But behind all this seeming arbitrariness lies a vast and ancient fabric of choices that are more (or more obviously) rational. Our survival as a species depended on people knowing how to make fire; that it burns but also cooks; that these berries are edible and those poisonous; that you can fish from this river but that those woods are best avoided, and so on. It follows from the fact that we exist at all that our ancestors acted mostly on experiences that told them these things and not on whether A is a Pisces or B dreamt of seven fat cows.

It took longer to develop a scientific basis for decision-making that applied not only when the relevant facts were known but also when they were uncertain. If you know green berries are poisonous, you shouldn't eat them. What if you know green *or* red berries are poisonous but can't remember which? A systematic approach to such cases had to await two surprisingly late discoveries.

One was probability. Probability applies most simply in some games of chance. It is clear enough what it means to say that the probability of dealing the ace and king of hearts in poker is about 1 in 332. Here we take the probability of an event to control or to arise from the frequency of other events that resemble it in some obvious way. To find the probability of, for example, 'ace and king of hearts' on this deal you look at how often they turn up in other deals.

But even many 'games of chance', like betting on a horse, turn on events that don't naturally fall into a large class of similar events. If in 2011 you wanted to know the probability that Red Cadeaux wins the Melbourne Cup, you would look at – what? How often he won it before? But he never ran it before. How often he wins against Dunaden, who is also racing? But he never ran against Dunaden. How often he wins any race at all? But he ran those other races in widely varying conditions against widely varying opposition, and so on.

[1] Kahneman and Tversky 1984: 347 (so note that these are 1980s dollars). [2] Seaman 2002.
[3] Speck 1935 ch. VI. [4] Livy *Ab Urbe Condita* 1.7.

But there is such a thing as *subjective* probability, or confidence. Your confidence in an event needn't depend on (or settle) how often anything similar happens – you might be ignorant of that. Of course, it *may* so depend; but the point is that we can measure confidence without being dogmatic about that. In a way people have known what confidence is for as long as they have felt it; but we can trace its 'discovery' to Frank Ramsey's famous paper of 1926, which also tells you how to measure it.[5]

The second discovery was utility or subjective value. The idea that some things have 'real' value, independently of what any one person thinks or wants, was central to Plato's philosophy and doubtless part of the interior decor for centuries before. *Price* is a kind of objective value, if not what Plato had in mind. The price of something may depend (in some market conditions) on the totality of people's wants, but it doesn't depend on any one person's wants. Similarly, evolutionary fitness – propensity to reproduce – is an objective value, at least wherever fitness is independent of anyone's opinions or tastes.

But what matters for decision-making is not objective value but what the decision maker wants. Diamonds cost more than water, but you wouldn't care if you were thirsty. Everyone knows that some religions implicitly or explicitly encourage their followers to reproduce, but nobody joins them for that reason. What motivates you and me is *subjective* value: what you want and how much you want it. People have known in a way what subjective value is for as long as they have wanted things. But we can trace *its* 'discovery' to Daniel Bernoulli's famous paper of 1738 on how to measure it.[6]

Normative decision theory arises from the interplay of subjective probability and subjective value. It is like a machine with inputs and outputs. Suppose that you are facing a set of options and you don't know what to do. Then the inputs to normative decision theory are (a) what you think (i.e. subjective probability); (b) what you want (i.e. subjective value). And the *output* is a recommendation from the options.[7]

For instance, suppose I must bet $1 on Dunaden or on Red Cadeaux. If I win on Dunaden, I make 25¢. My subjective value for this outcome is +10. If I win on Red Cadeaux, I make 10¢. My subjective value for that is +8. If I lose on either, I lose $1. My subjective value for that is zero. I'm 25% confident that Dunaden will win and 75% confident that Red Cadeaux will.

My options, the possible results of the race, my confidence in the latter and my values for the resulting outcomes are as in this table.

---

[5] Ramsey 1926. De Finetti 1937 is an independent treatment on similar lines.

[6] Bernoulli 1738.

[7] For an extended introduction to normative and other forms of decision theory see Peterson 2017.

Table 1.1 Horse race

	RC wins	**D wins**
Bet on RC	+8, 75%	0, 25%
Bet on D	0, 75%	+10, 25%

In each cell there are two numbers: first, the *value* of the corresponding outcome if I take the corresponding option; second, my *confidence* in the outcome if I take the option. For example, the top left-hand cell (+8, 75%) says (a) that I am 75% confident that if I bet on RC, then RC wins; (b) that this is worth +8 to me. Similarly with the other three entries. Now, how should I bet?

A simple approach calculates the *expected value* of each option. For each option this is got by adding the subjective value of each outcome if you choose that option, multiplied by the subjective probability of that outcome if you choose that option. One simple normative decision theory then says: choose any option with the highest expected value. I'll call this theory **MEU** ('Maximize Expected Utility').

Thus in Table 1.1 the expected value of a bet on Red Cadeaux is 6. The expected value of a bet on Dunaden is 2.5. So MEU advises betting $1 on Red Cadeaux.[8]

MEU is one of many normative theories. There is a theory that advises you to choose the (or any) option whose best possible outcome you like most ('maximax'): here, a bet on Dunaden. There is a theory that advises you to choose any option whose *worst* possible outcome you like most ('maximin'): this theory finds both bets acceptable. There are many others.

But something like MEU is appealing. The connection between value and expectation is a consequence of plausible assumptions.[9] And generalizing addition and multiplication in various natural ways reveals a correspondingly generalized idea of expectation within many approaches to decision-making.[10] The theory is simple and gives correct advice where the right decision is obvious.

One version of MEU is the subject of this Element: *Evidential Decision Theory* or EDT. EDT is the normative theory which (according to me) gets choice right: given what you think and want, it gives rational advice about what to do.

[8] Expected value for a bet on RC is 8(75%) + 0(25%) = 6; for a bet on D: 0(75%) + 10(25%) = 2.5.
[9] Milne and Oddie 1991: 54–8. For the merits of MEU-style 'linear pooling' see Pettigrew 2019 ch. 9.
[10] Chu and Halpern 2004.

But what this involves is stranger and more austere than you might expect. Most people think that any general account of how to behave ought to mention the *effects* of your behaviour: what it causes or brings about. But EDT has no special place for that relation. What matters about an option is what it *indicates* – whether by bringing it about or by being symptomatic of it in other ways. As we'll see, this has unsettling practical and philosophical consequences.

1.2 What It Is Not

Before getting into all that, I should distinguish my topic from two others.

First: descriptive decision theory. In one way, a descriptive theory is like a normative theory: it takes beliefs and desires as inputs and gives options as outputs. The difference is in what it is *meant* to do: the normative theory tells you what to do, but the descriptive theory is supposed to predict what you will, in fact, do. Normative MEU theory advises you to maximize expected utility; but descriptive MEU predicts that you will. Given, for example, beliefs and value as in Table 1.1, descriptive MEU predicts that you *will in fact* bet on RC (but it doesn't say that you *should*).

Descriptive decision theory belongs to ethology, whereas normative decision theory belongs to ethics. Facts about actual behaviour might refute some descriptive decision theory but not its normative counterpart. For instance, the well-known 'paradoxes' of Allais and Ellsberg seem to refute *descriptive* MEU. They present situations where subjects apparently make choices that are not maximizing the expectation of anything.[11] But they don't refute *normative* MEU, not if the Allais and Ellsberg subjects are behaving irrationally. And this combination – accepting normative but rejecting descriptive MEU – was a popular reaction to Allais's findings.[12] In any case, EDT itself has descriptive and normative versions. The normative interpretation takes centre stage here. The point is not to describe your behaviour but to guide it.[13]

To explain the second thing I won't discuss, I distinguish *behaviouristic* from *psychological* decision theory.

Behaviouristic decision theory states principles of predicted or recommended behaviour that don't mention anything mental – what you think or want. They just interrelate choices. One such principle is 'transitivity of preference': if you'd choose A over B ('you prefer A to B'), and B over C ('you prefer B to C'), then you'd choose A over C ('you prefer A to C'). 'Preference' here is just behaviour: preferring A to B, on this reading, *means* being disposed to choose A when B is the alternative.

---

[11] Allais 1953; Ellsberg 1961. [12] Moscati 2019: 190.

[13] EDT has potential as a descriptive theory: see Grafstein 1991, 1999.

Psychological decision theory specifies behaviour as a function of explicitly psychological parameters. MEU is a psychological decision theory. It specifies behaviour as a function of what you think (subjective probability) and what you want (subjective value).

The behaviouristic/psychological and normative/descriptive distinctions create four possibilities for decision theory:

- behaviouristic and normative
- behaviouristic and descriptive
- psychological and normative
- psychological and descriptive.

Psychological normative and psychological descriptive decision theories include the readings of MEU described above. The behaviouristic principle of transitivity might be understood descriptively: people do, in fact, typically behave this way: if a person chooses A over B and B over C, then she will, in fact, also choose A over C. Or it might be understood normatively: if you choose A over B, and B over C, but *not* A over C, then you are choosing irrationally.

Table 1.2 lists principles illustrating all four kinds of theory. For instance, the entries in the top row are behaviouristic: neither specifies what you think or want. But the top-left entry *prescribes* behaviour, whereas the top-right entry *predicts* it.

Given a behaviouristic theory *B* and a psychological theory *P*, there may be a *representation theorem* connecting them. This says that *if* your behaviour conforms to *B*, then we could simulate ('represent') your behaviour by

Table 1.2 Four kinds of decision theory

	Normative	**Descriptive**
Behaviouristic	If you choose apples over pears, and pears over bananas, then you should choose apples over bananas.	If you choose apples over pears, and pears over bananas, then you will choose apples over bananas.
Psychological	If you think apples more nourishing than bananas and only want nourishing food, then you should choose apples over bananas.	If you think apples more nourishing than bananas and only want nourishing food, then you will choose apples over bananas.

programming any of a suitable range *m* of mental states into someone who conformed to *P*. For instance, Savage showed that anyone whose choices satisfied his behaviouristic theory could be represented as having beliefs and desires from a given range and conforming to a version of MEU.

Representation theorems are contributions to philosophy of mind. We can speculate about whether dogs or spiders (or plants or cricket bats) are conscious. But simple behaviour makes it empirically pointless to postulate a complex mentality to things that behave simply. 'We say a dog is afraid his master will beat him; but not: he is afraid his master will beat him tomorrow. Why not?'[14] Because the dog's behaviour is not so complex that any explanation of it would have to distinguish thoughts about tomorrow from thoughts about today.

Representation theorems say precisely what patterns of behaviour *would* give empirical point to attributing this or that belief-desire mentality to the thing doing the behaving. And by specifying what *range* of attributions would explain the behaviour, they also say *how much* mentality it compels us to attribute. This is philosophically interesting from any perspective. Given even moderate behaviouristic sympathies, it takes on special importance: it tells us what it takes to have a mind.[15]

Standard expositions of decision theory typically offer (a) a behaviouristic theory, (b) a psychological theory, and (c) a representation theorem. Jeffrey's classic exposition of Evidential Decision Theory involved (a)–(c). And much of the mathematical and philosophical ingenuity in his and in Bolker's work lay in their discovery of the behaviouristic axioms and the representation theorem.[16] Because of its philosophical importance and interest, Appendix C tries to spell out the intuition behind (c), the representation theorem.

But for the main part I focus on (b). That is, this Element mainly concerns EDT considered as a *normative psychological* thesis. If you tell it what you think and want, it tells you what to do.

1.3 Plan of This Element

Section 1 explains subjective probability – how confident you are that something is true, and subjective or *news* value – how much you want it to be true. Then I introduce Evidential Decision Theory, which recommends maximizing

[14] Wittgenstein 1953: §650.

[15] An analogy from philosophy of language: Quine's argument for inscrutability of reference is a representation theorem connecting linguistic behaviour with the theory assigning references to the speaker's terms (Quine 1981: 19–20). Quine shows that different assignments represent the speaker's behaviour equally well. He infers that there is no fact about which assignment is correct. As we'll see, the Bolker–Jeffrey representation theorem may not get us this far, because it is dubious just how behaviouristic their 'behaviouristic axioms' are. See Appendix C, Section 3.1.

[16] Bolker 1966, 1967; Jeffrey 1983.

news value. Then I sketch the orthodoxy to which EDT is a challenge: Causal Decision Theory.

Sections 2–4 display the content of EDT via its unorthodox recommendations. Each section highlights a different feature of EDT:

- It recommends options that signal good news but do nothing to cause it.
- It recommends options that scramble their own signals of bad news.
- It evaluates future options in the same way as present ones.

In all these cases I find EDT defensible, each section arguing briefly to that effect. But their main aim is less to convince you that EDT is true than to make vivid what it means.

What emerges is not just advice but a vision of decision-making. Your choices don't flow from some part of you that has, or that you for some reason believe to have, a power to intervene in the external world without itself being subject to that world. On the contrary, your decision-making processes are as subject to external influence as your digestive processes. Strange to say, only Evidential Decision Theory fully accommodates this fact.

2 News Value

Evidential Decision Theory says, 'Do what you most want to learn that you will do.'

It can be stated and can sometimes be applied without specifying any measure of *how much* you want to learn something. But the clearest way to explain it is via a measurable quantity called *news value*, the background to which the next two sections explain.

2.1 Possible Worlds and Propositions

To model decision-making under uncertainty, we must represent what you are uncertain about. That means considering possible ways things might be, given what you know when choosing. For example, if you are betting on horses, there is a possible situation where you bet $1 on Kelso and he wins, another where you bet $30 on Trigger and Swaps wins, and so on.

Possible worlds (sometimes 'worlds') are what I'll call the possible ways things might be. Each world settles everything: by choosing which world to realize, God settled all of history. There is then a vast, maybe infinite range of possible worlds, one for each possible history. I'll mostly treat worlds as mathematical points, but nothing important is lost – and some vividness is gained – by imagining them as concrete universes, spatio-temporally isolated

Figure 2.1 The set of possible worlds

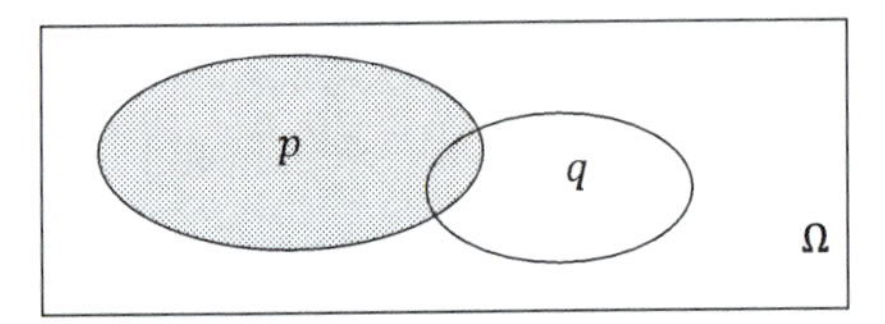

Figure 2.2 The propositions p and q

from ours, at which these histories really occur.[17] Here I'll write @ for our possible world (the 'actual world'), and $\boldsymbol{\Omega}$ for the set of all possible worlds.

Figure 2.1 represents Ω and the points and regions within it. The rectangle is the set Ω of worlds (i.e. the set of points inside it). One of these points is @, the actual world.

Every *set* of worlds corresponds to some condition that @ may meet or fail to meet. If p is the set of all worlds where it rains in Tokyo on New Year's Day 2020, p corresponds to the condition that it rains in Tokyo on New Year's Day 2020. If q is the set of worlds where somebody one day runs 100 m in less than 9.5 seconds, then q corresponds to the condition that somebody one day runs 100 m in less than 9.5 seconds. Any such set is a **proposition**. Each such set corresponds to some set of points in the rectangle. I'll indicate these sets as shaded regions of Ω: see Figure 2.2.

I'll use concisely stateable conditions to label the corresponding sets. For instance, p is the proposition *that it rains in Tokyo on New Year's Day 2020*. A proposition p is **true at a world** w if w meets the corresponding condition: that is, if w belongs to the set p, written $w \in p$. Otherwise p is **false at** w. A proposition is **true** or **false** simpliciter if it is true or false at @.

The familiar set-theoretic operations on propositions correspond to logical operations in the obvious way. For propositions p and q:

- $\boldsymbol{p \cup q}$ ('p or q') is the proposition that is true at a world w if and only if either p is true at w *or* q is(or both).

[17] Following Lewis 1986.

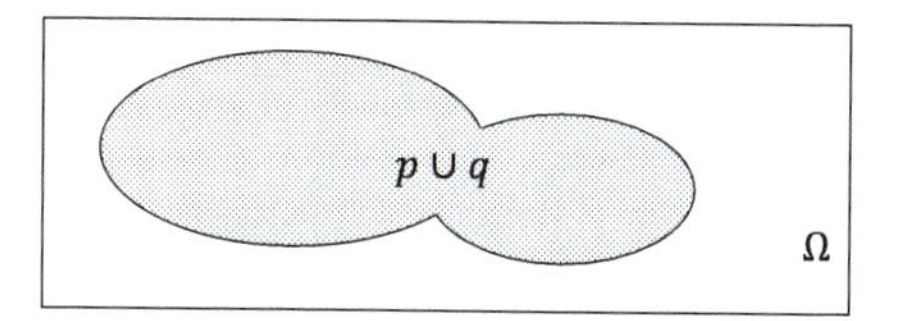

Figure 2.3 The proposition *p* or *q*

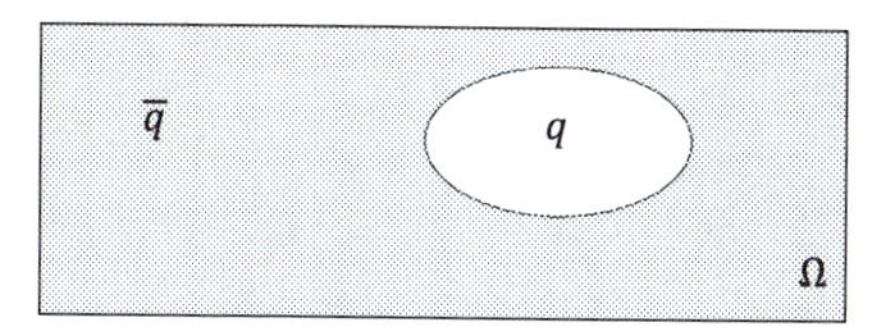

Figure 2.4 The proposition *not q*

- $\boldsymbol{p \cap q}$ (' *p* and *q*') is the proposition that is true at *w* if and only if *p* is true at *w* *and q* is.
- $\boldsymbol{\overline{p}} = \Omega - p$ ('not *p*') is the proposition that is true if and only if *p* is false at *w*.

In our example, $p \cup q$ is the proposition that *either* it rained in Tokyo on New Year's Day 2020 *or* somebody one day runs 100 m in less than 9.5 s: see Figure 2.3. And $\overline{q}$ is the proposition that nobody ever did or will run 100 m in less than 9.5 s: see Figure 2.4.

It may happen that *p* and *q* are never true at the same world. For instance, let *p* be the proposition that Tiger Roll wins the Grand National in 2019 and *q* the proposition that Magic of Light wins it. In that case they are **incompatible**. And $p \cap q$ is the set with *no* elements, the **empty set** $\varnothing$.

A **partition** of Ω, finally, is a set of non-empty propositions $\{q_1, q_2 \ldots q_n\}$ such that every world lies in exactly one element or **cell** of the set.[18] For instance, if in all possible worlds the Grand National was won in 2019 by exactly one of the 40 starters, but it might have been any of them, then there is a partition $\{q_1, q_2 \ldots q_{40}\}$ where each cell corresponds to one of the starters: q_1 is the proposition that Tiger Roll won it, q_2 is the proposition that Magic of Light won it, and so on. See Figure 2.5.

2.2 Subjective Probability

So much for the objects of uncertainty. Now for its measurement. You may be more certain of one thing than another. I am highly confident that next July the

[18] Notwithstanding my notation, a partition may be infinite, though none of the applications here require this.

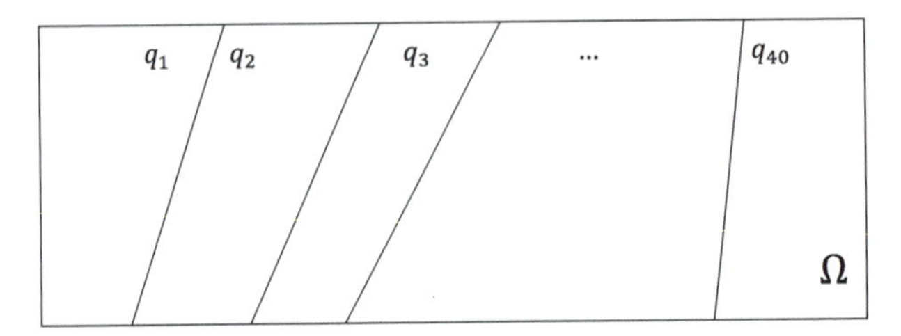

Figure 2.5 Partition of the set of possible worlds

average temperature (in the UK) will exceed what it was last December. I am less confident that it will snow next January; less confident still that it will snow next August. These comparisons suggest a scale for measuring confidence. It runs from zero to one and is called **credence** or **subjective probability**.

Intuitively we can visualize confidence in a proposition as the *area* of the corresponding region. See again Figure 2.2. Suppose you know that exactly one point in Figure 2.2 is the actual world, and you are equally confident, for any region of a given area, that *it* contains the actual world (as if God chooses which world to actualize by randomly sticking a pin in Ω). Then for any given region, your confidence that @ lies in *that* region is proportional to its area. So your confidence that a proposition is *true* (i.e. true at @) corresponds to the associated area.

For instance, your confidence that it rains in Tokyo on New Year's Day 2020 is the area of the shaded region in Figure 2.2. We can express this area as a proportion of the whole rectangle. The closer it is to 1, the greater your confidence that the proposition is true. The closer it is to 0, the greater your confidence that it is false.

More formally, credence is a **probability function**: it assigns to each set of worlds (proposition) p a number $Cr(p)$ between 0 and 1, such that the following rules hold for any propositions p and q:

$$Cr(p) \geq 0 \tag{2.1}$$

$$Cr(\Omega) = 1 \tag{2.2}$$

$$p \cap q = \varnothing \rightarrow Cr(p \cup q) = Cr(p) + Cr(q). \tag{2.3}$$

These rules make intuitive sense when we interpret probability as area. (2.1) says that every region has some non-negative area (maybe zero). (2.2) says that the area is measured in units of the area of the whole rectangle Ω. (2.3) says that

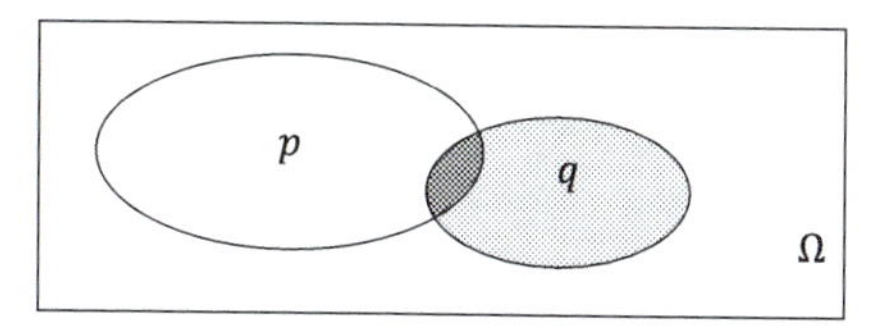

Figure 2.6 Conditional probability

probability is **finitely additive**: if two regions have no overlap, then the area of their composite is the sum of their areas.[19]

They also make sense if probability means confidence. (2.1) says that your confidence in any proposition is at least zero; (2.2) that your confidence that some world is actual is 1, or 100%: we can take these as definitions of the scale on which you measure confidence. (2.3) says that your confidence in a proposition that may hold in exactly one of two possible ways is the sum of your confidences in each of those ways. If you are 24% confident that (p) Tiger Roll will win the Grand National and 12% confident that (q) Magic of Light will win, then you are 24% + 12% = 36% confident that ($p \cup q$) one of them will win.

Next, we make an assumption about what happens when you *learn* something. Suppose you learn for certain some previously uncertain proposition q. This affects your credences: they will shift from the probability function Cr to a new probability function; call it Cr_q. The **Bayesian** assumption is that for any p, your new credence in p is the ratio of your old credence that it and q are *both* true to your old credence in q. This quantity is written $Cr(p|q)$ and called your **credence in** p **given** q. So the assumption is as follows:

$$\text{If } Cr(q) > 0, \text{ then } Cr_q(p) = Cr(p|q) =_{\text{def.}} \frac{Cr(p \cap q)}{Cr(q)}. \tag{2.4}$$

Pictorially, $Cr(p|q)$ is the proportion of the q region that lies in the p region. If you learn that @ lies in the q region (the lightly shaded region in Figure 2.6), you can discard the entire $\overline{q}$ region. Your confidence that @ also belongs to the p region must be the proportion of the remaining area that still belongs to the p region (the heavily shaded region).

For instance, suppose you start out 24% confident that (p) Tiger Roll will win and 12% confident that (q) Magic of Light will. Now you learn that ($p \cup q$) *one of them will* win. Then your new credence that Tiger Roll will win is $Cr(p|p \cup q) = \frac{24\%}{36\%} = 66.7\%$. Your new credence that Magic of Light will win is $Cr(q|p \cup q) = \frac{12\%}{36\%} = 33.3\%$. That is the Bayesian assumption about learning.

[19] (2.3) here replaces the usual *countable* additivity axiom. This simplification won't affect the following discussion.

The exposition so far is not ambitious, not compared to Bolker's and Jeffrey's. Their representation theorem *justifies* assigning credences and values to anyone whose preferences satisfy their behavioural axioms (see §1.2). Nothing here does any of that. Nor did I justify my claim that (2.4) describes how those beliefs change when you learn something new. I just took all that for granted. But doing so supplies the material for a simple explanation of news value and then of EDT itself.

2.3 Binary News Value

As well as believing things with varying confidence, people also want them with varying intensity. Just as subjective probability measures the conviction with which you believe something to be true, news value measures the intensity with which you want it to be true.

Imagine Ω again as a plane of which some region g is grey and the rest white (see Figure 2.7). And imagine that you care about just one thing: that @ is grey (i.e. in g). More explicitly, imagine a proposition g whose truth or falsity settles *everything* that matters to you. g might describe your pre-election by God if you are a Calvinist, or the extermination of the French nobility if you are Madame Defarge. Or it might be anything else: the point is that its truth or falsity exhausts everything you care about. Once you know whether g is true or false, nothing else matters; and what you really want is that g is true.

(It is, of course, no accident that my examples are paradigms of single-mindedness. Most people do *not* care about just one thing. But pretending that they do makes life simpler. Appendix A shows how to dispense with this fiction.)

If such a g exists, we can use it to measure the value to you of learning any *other* proposition. Letting p be any such proposition, the value of learning p is just this: how confident you would *then* be of getting your heart's desire, that is, how confident you would then be that g is true. What this quantity is, follows from the Bayesian assumption (2.4): it is $Cr_p(g) = Cr(g|p)$. We therefore define the **news value** of a proposition p to be:

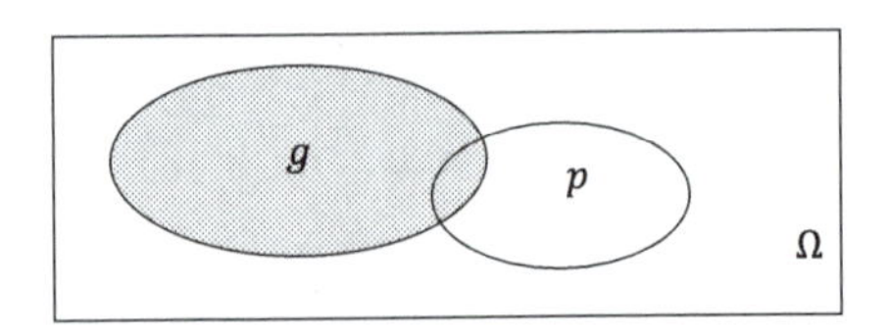

Figure 2.7 The proposition g

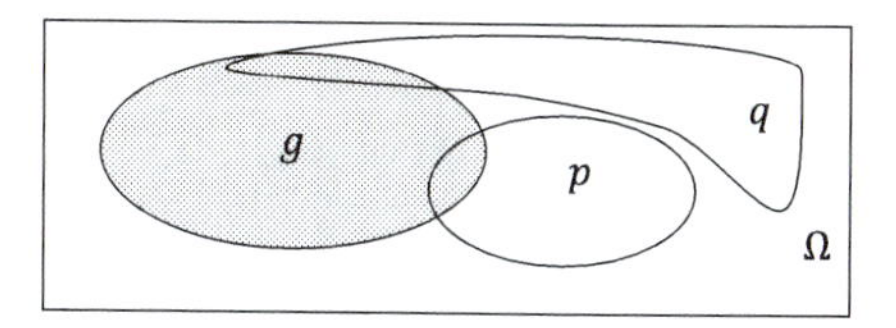

Figure 2.8 Two non-overlapping regions that overlap with g

$$V(p) = Cr(g|p) = \frac{Cr(g \cap p)}{Cr(p)}. \tag{2.5}$$

The news value of p is undefined when $Cr(p) = 0$.[20]

We can now use this definition to derive a formula which will be useful when it comes to calculating V in actual cases.

If p is any region, write $\boldsymbol{\Gamma(p)}$ for the area (credence) of the grey region within it:

$$\Gamma(p) =_{\text{def.}} Cr(g \cap p).$$

Clearly Γ, like Cr itself, is finitely additive (cf. 2.3): the area of grey within the union of any two non-overlapping regions is the sum of the areas of grey in each. Figure 2.8 displays two non-overlapping regions p and q such that $\Gamma(p \cup q) = \Gamma(p) + \Gamma(q)$.

[20] Those familiar with *The Logic of Decision* will note that here (following Lewis 1996) I depart from Jeffrey's convention setting the news value of Ω at zero for any Cr (Jeffrey 1983: 99, putting T for Ω). That convention is inconsistent with $V(p) = Cr(g|p)$, which implies $V(\Omega) = Cr(g) > 0$ unless all propositions have the same news value. In fact, on the present theory, the news value of Ω changes with one's confidence in g. This follows from the fact that (2.5) entails **invariance**, the thesis that the news value of a proposition is independent of whether you think it is true. Formally, $V_p(p) = V(p)$ This is the news value that you would have for p if you were to learn that p. Bradley and Stefánsson argue that invariance is false because once you have just learnt p, p should not *then* be especially good (or bad) news, even if it was just *before* you learnt it (2016: 700–2). This is true, but invariance does not entail that what you have just learnt ever *is* especially good (or bad) news just after you've learnt it. On the contrary, (2.5) entails that any pleasant surprise – however mildly pleasant, and however mildly surprising – is *better* news than p if you have just learnt p. For if I consider q to be both a surprise and good news given p, then learning that p is true automatically makes q better news than p. Proof: the 'surprise' bit implies (a) $Cr(q|p) < 1$. The 'good news' bit implies $V_p(q) > V_p(\overline{q})$, hence by (2.5) we have $V(p \cap q) > V(p \cap \overline{q})$. But it also follows from (2.5) and the rules of probability (2.1)-(2.4) that $V(p) = Cr(g|p) = Cr(g|p \cap q)Cr(q|p) + Cr(g|p \cap \overline{q})Cr(\overline{q}|p) = V(p \cap q)Cr(q|p) + V(p \cap \overline{q})Cr(\overline{q}|p)$: that is, (c) $V(p) = V(p \cap q)Cr(q|p) + V(p \cap \overline{q})Cr(\overline{q}|p)$. But (a)-(c) implies $V(p \cap q) > V(p)$, hence by (2.5) $V_p(q) > V(p)$ and therefore by invariance, which follows from (2.5), we have $V_p(q) > V_p(p)$.

By a similar argument, (2.5) also entails that any unpleasant surprise is *worse* news than anything that you have just learnt. Invariance does therefore secure the intuitive *neutrality* of p after you have learnt it.

Now suppose God divides the plane exclusively and exhaustively amongst his angels. There are n angels: he gives region q_1 to Angel 1, q_2 to Angel 2 …, q_n to Angel n, where $q_1, q_2 \ldots q_n$ form a partition (as in Figure 2.5). Any *other* region p must therefore be made up of exactly these non-overlapping regions: the subregion of p owned by Angel 1, the subregion of p owned by Angel 2 … the subregion of p owned by Angel n. More formally:

$$p = \bigcup_{i=1}^{n} p \cap q_i.$$

Since these regions don't overlap, it follows from the additivity of Γ that the grey region in any p is just the sum of the grey regions in the subregions of p owned by each angel:

$$\Gamma(p) = \sum_{i=1}^{n} \Gamma(p \cap q_i). \tag{2.6}$$

We defined the news value of a proposition as the proportion of the corresponding region that is grey, that is, the ratio of the measures Γ and Cr (see (2.5)). Equivalently, the area of grey within any region r is just the product of its area and the proportion of it that is grey, that is:

$$V(r) = \frac{\Gamma(r)}{Cr(r)}$$

$$\Gamma(r) = Cr(r)V(r).$$

Applying this to (2.6) and assuming each $Cr(p \cap q_i) > 0$, we have:

$$Cr(p)V(p) = \sum_{i=1}^{n} Cr(p \cap q_i)V(p \cap q_i).$$

The definition of conditional probability (2.4) now implies the fundamental equation interrelating news value and subjective probability. It says that if p is any proposition and if $\{q_1, q_2 \ldots q_n\}$ is any partition such that $Cr(p \cap q_i) > 0$ for each q_i, then:

$$V(p) = \sum_{i=1}^{n} Cr(q_i|p)V(p \cap q_i). \tag{2.7}$$

That is, the news value of p is a weighted sum of the news values of all the ways p might hold, where the weight attached to each way is your confidence that p holds in that way given that it holds in *some* way.[21] I call this equation

[21] For a formal proof of (2.7): $V(p) = Cr(g|p) = \sum_{i=1}^{n} Cr(g \cap q_i|p)$ by (2.3) and (2.5). But $\sum_{i=1}^{n} Cr(g \cap q_i|p) = \sum_{i=1}^{n} \frac{Cr(g \cap p \cap q_i)}{Cr(p \cap q_i)} \cdot \frac{Cr(p \cap q_i)}{Cr(p)} = \sum_{i=1}^{n} Cr(q_i|p)V(p \cap q_i)$ by (2.3) and (2.5) again.

'fundamental' not because it is axiomatic but because it is (as we'll see) endlessly useful.

This basic story gives enough grip on news value for the applications and philosophical arguments to come. Two obvious refinements are available: for details see Appendix A. But since here it makes no difference except to simplify things, I'll keep the (false) assumption that all you ultimately care about is g. Accordingly the news value V of a proposition p will mean what (2.5) says, expressed for convenience in percentage terms. I write, for example, $V(p) = 95$ to abbreviate $V(p) = 95\%$, or equivalently $Cr(g|p) = 0.95$.

2.4 Examples

It will help to consider some examples.

> *Cheltenham Gold Cup*. You are about to learn whether (p) it rains tonight or ($\overline{p}$) not. Tomorrow is the race. You have put your last \$10 on Al Boum Photo (ABP) at odds of 1–2. You know that if the going is hard, firm, or good (and it will be if tonight is *dry*), then ABP has a good chance of winning: your confidence in his winning, on learning that it will be dry, would be 70%. But if the going is soft or heavy (which it will be if it rains tonight), then your confidence that ABP will win is only 20%. All that matters about the race is how much you win.

What are the news values of p and of $\overline{p}$?

As far as you care, there are two possible results: (q) ABP wins, and ($\overline{q}$) he does not. A win for ABP is worth \$5. If ABP doesn't win, you lose \$10. Suppose that at levels of wealth close to yours, every dollar gained or lost increases or decreases your confidence in g by two points (2%); set your initial confidence in g at 50%. So:

$$V(q) = 50 + (5 \times 2) = 60.$$

$$V(\overline{q}) = 50 - (10 \times 2) = 30.$$

Knowing the result would make you indifferent to the weather last night. The news that ABP wins following rain tonight has as much value as the news that he wins following no rain, both being worth \$5. Similarly, news that he doesn't win is equally unwelcome, whatever tonight's weather. So:

$$V(p \cap q) = V(\overline{p} \cap q) = V(q) = 60$$

$$V(p \cap \overline{q}) = V(\overline{p} \cap \overline{q}) = V(\overline{q}) = 30.$$

Now subjective probabilities. These follow straightforwardly from the description.

$$Cr(q|p) = 0.2$$

$$Cr(\overline{q}|p) = 0.8$$

$$Cr(q|\overline{p}) = 0.7$$

$$Cr(\overline{q}|\overline{p}) = 0.3.$$

Substituting these into (2.7), with $\{q, \overline{q}\}$ as the partition $\{q_1 \ldots q_n\}$, yields:

$$V(p) = 36$$

$$V(\overline{p}) = 51.$$

That it rains tonight is bad news: as bad as learning that you have lost \$7. That it does not rain is moderately good news: as good as learning that you have won 50¢.

This is intuitive. After all, rain gives the better outcome a probability of only 20%. So its news value should be 20% of the way along the line from the value of the worse outcome to the value of the better outcome. Similarly, no rain gives the better outcome a probability of 70%. So its news value is 70% of the way along the same line.

Now a second example.

> *Luke* v. *Emperor*. Just before the final showdown, Luke wonders – did I replace the battery in my lightsabre? He can't remember. Batteries cost \$1, but Luke's chances of defeating the Emperor with a new battery are 60%; without it, only 10%. And it matters that he defeats the Emperor: if he doesn't, the universe will be plunged into a millennium of tyranny and bloodshed. Avoiding that would for Luke personally be as good as winning \$10, and incurring it as bad as losing \$10.

Suppose again that every dollar gained/lost increases/decreases Luke's confidence in g by two points; set his initial confidence in g at 50%. These tables summarize his position.

Each entry in Table 2.1 gives the news value to Luke of the outcome that combines the corresponding row and column headings. For instance, the entry for p and q, where Luke defeats the Emperor after having replaced the battery, is 68. It is as good as winning \$9: \$10 for saving the universe minus \$1 for the new battery.

In Table 2.2 each entry gives Luke's confidence in the column heading, given the row heading. For instance, the entry corresponding to $\overline{p}$ and q is 0.1, because learning that he did not replace the battery gives Luke 10% confidence that he wins: $Cr(q|\overline{p}) = 10\%$.

Table 2.1 Values of outcomes in *Luke* v. *Emperor*

	q: **Luke wins**	$\overline{q}$: **Emperor wins**
p: **new battery**	68	28
$\overline{p}$: **old battery**	70	30

Table 2.2 Probabilities of outcomes in *Luke* v. *Emperor*

	q: **Luke wins**	$\overline{q}$: **Emperor wins**
p: **new battery**	0.6	0.4
$\overline{p}$: **old battery**	0.1	0.9

Apply (2.7) to calculate news values for p and $\overline{p}$:

$$V(p) = (0.6 \times 68) + (0.4 \times 28) = 52$$

$$V(\overline{p}) = (0.1 \times 70) + (0.9 \times 30) = 34.$$

Luke prefers to learn that he replaced the battery.

But notice that although the news that Luke replaced the battery is better than the news that he didn't, the latter still *dominates* the former over the partition $\{q, \overline{q}\}$. I mean (i) given q is true, $\overline{p}$ is better news than p: Luke would rather win with an old battery than with a new one because of the \$1 saving; (ii) but also, given $\overline{q}$ is true, $\overline{p}$ is *still* better news than p: Luke would rather lose with an old battery than with a new one, for the same reason.

So is Luke better off with an old battery? No, because which of p and $\overline{p}$ is true has a bearing on which of q and $\overline{q}$ is true. p makes the highly desirable q more likely than $\overline{p}$ does. The desirability of q in any circumstances exceeds that of $\overline{q}$ in any circumstances, and by enough that p's being a strong indicator of q easily compensates for its relative cost.

The example shows that when two different possible new items p and p' have different bearings on a variable of interest, p may be better news than p' even if, given any particular value of that variable, you'd find p' better news than p.[22]

[22] There is an obvious connection with Simpson's Paradox (Simpson 1951). For a given finite population Ω of which X is any subclass, write $\Pr(X)$ for the relative frequency with which members of Ω fall under X, so that Pr is a probability function. Then if there are classes X, Y, Z and a partition Π of Ω such that $\Pr(Z|X) > \Pr(Z|Y)$ but $\Pr(Z|Y \cap W) > \Pr(Z|X \cap W)$ for any $W \in \Pi$, these classes exhibit a Simpson-paradoxical statistical pattern. Putting $\Pi = \{q, \overline{q}\}$, $X = p$, $Y = \overline{p}$, and $Z = g$ and replacing *Pr* with *Cr*, this is what happens in *Luke* v. *Emperor*.

This account of credence and news value says nothing about action, choice, decision, and so on. That is unsurprising. There is no obvious absurdity (although there may be a subtle absurdity) in the idea of creatures with hopes and expectations but no agency – intelligent trees – whose attitude to external events is passive but not disinterested. They could have credences and news values too. What makes EDT a theory of *decision* is that it takes the news value of a proposition to settle not only its welcomeness as news but also its choice-worthiness as an option. To explain this, I turn to contexts in which propositions are both subjects of news and objects of choice.

2.5 Decision Problems

A decision problem is a situation where you choose from two or more options. It has four components:

- All the ways things could turn out: Ω
- What you think: credence function Cr
- What you want: news value function V
- Your options: a set $O = \{o_1, o_2 \dots o_n\}$, $n \geq 2$ where O is a partition of Ω, $Cr(o_i) > 0$ for $o_i \in O$.

A **decision problem** is formally a vector $\langle \Omega, Cr, V, O \rangle$ of these components. I have already discussed Ω, Cr and V. In principle, O can be any set of propositions of which exactly one is true, but you can't rule out any for sure. But for application its cells should be propositions that you can make true by *choice*.

'Choice' needn't mean anything metaphysical. EDT makes almost no commitments about the nature of choice. It says: give me your theory of what choice *is* and I'll tell you which choices are rational.

For instance, choice need not involve indeterminism. Your options may not *all* be consistent with the past and the laws of nature: you might know that only *one* of them is (but not which one). A proposition then describes an option for you if it would have been caused in the right kind of way (whatever that is) by suitably different beliefs and desires, even if the latter could not have been other than they actually are given the actual past and laws. In short, options are propositions that you 'make true as you please'.[23]

But choice does involve subjective *uncertainty*. Each option must be such that you are not *certain* that you will *not* realize it. This is because news value $V(o_i)$ is defined only if $Cr(o_i) > 0$. This poses no serious difficulty: all our examples

[23] Jeffrey 1983: 84.

concern agents who are deliberating between options and cannot rule any out for certain.[24]

It is implicit in this framework that the objects of choice are objects of news value and uncertainty: propositions themselves. What one chooses are not, for example, bodily movements but propositions describing them. I do not choose between walking and driving to work but between the proposition that I walk to work and the proposition that I drive.

It sounds odd to say that one chooses abstract objects like propositions, but it does no harm. Whatever the objects of choice really are, we can apply EDT to propositions given a one-to-one correspondence connecting propositions and 'real' options. We apply the correspondence to generate propositions from 'real' options, apply EDT to these propositions to yield the rational (propositional) options, then reverse the correspondence to get the rational 'real' options. This mirrors the application of arithmetic to (abstract) numbers to count or measure physical objects.[25]

A subtler concern is that propositions cut too fine. This morning, Bernard J. Ortcutt thought he had two options for getting to work: take the next bus owned by Routemaster Co., or take the next bus owned by Stagecoach Co. Unknown to him, Routemaster and Stagecoach had merged overnight: the next bus owned by Routemaster *was* the next bus owned by Stagecoach. So he had one option but falsely thought he had two. And counting by propositions gives two: 'Ortcutt takes the next Routemaster bus' and 'Ortcutt takes the next Stagecoach bus' express different propositions even if 'Routemaster' and 'Stagecoach' happen to denote the same operator.

Reply: the options on our definition must form a partition: it should be impossible for two to be true at once. But 'Ortcutt takes the next bus owned by the Routemaster Co.' and 'Ortcutt takes the next bus owned by the Stagecoach Co.' *are* both true at once. So those sentences don't exhaust Ralph's options: if he thinks they do because he is certain that Routemaster and Stagecoach are different operators, then the error lies not with our definition of his options but with his own beliefs about them.

---

[24] One objection is that there are cases where an option is so clearly absurd that you are (or should be) certain that you won't realize it; but still it remains an *option*: in a choice between (o_A) drinking water, (o_B) drinking orange juice, and (o_C) drinking paint, o_C remains an option even if you are certain that you won't realize it. But (a) this problem goes away if your confidence in o_C is positive, however small it is; and it is implausible that you really are *completely* certain of not slipping. (b) Waiving (a), there is no harm to the theory in denying that o_C is an option after all: given a set O of options of which those in some subset O^* are certainly unchosen because absurd, the rational choices from the remaining set $O - O^*$ must be exactly the rational choices from O given the choice-theoretic principles α and β (Sen 1971). For further discussion, see Jeffrey 1977.

[25] This is the 'representational' theory of measurement: see e.g. Luce and Suppes 2004.

Ortcutt does *have* options. There is some partition of Ω into cells of which he can make true whichever he likes. It is $O^* = \{o_1, o_2, \ldots\}$, where o_1 says that he takes the next bus to come along, o_2 that he takes the second bus to come along, and so on. We can model his decision problem using O^*. Ortcutt still *thinks* that no option involves taking a Routemaster *and* a Stagecoach bus. But then, everyone holds some false beliefs; and his holding this one doesn't stop us advising him what to do.

2.6 Evidential Decision Theory

Given a decision problem, which option should I take? **Evidential Decision Theory** advises the option that maximizes news value (i.e. V); if more than one does, it permits any that does. It says: do what you most want to learn that you will do.

An example:

> *Driving Test.* My driving test is next week. Passing is worth \$20 to me. I can choose now whether to book a lesson. The lesson costs \$5. I know that 80% of people who take a lesson in advance of the test pass, whereas only 25% pass without a lesson. Should I pay for the lesson?

Here Ω covers all possibilities, Cr and V reflect my beliefs and desires as described below, and my option set is $O = \{o_1, o_2\}$, where o_1 says that I pay for the lesson, o_2 that I don't.

Suppose these statistics determine my confidence of passing or failing the test given that I have, and given that I haven't, taken the lesson. Suppose my initial confidence in g is 50%; and each additional dollar gained (or lost) increases (or reduces) that confidence by 1 point. Let s_1 say that I pass and s_2 that I fail. So the values and probabilities of outcomes are as in Table 2.3 and Table 2.4:

Table 2.3 Values of outcomes in *Driving Test*

	s_1: **pass**	s_2: **fail**
o_1: **pay for lesson**	65	45
o_2: **don't**	70	50

Table 2.4 Probabilities of outcomes in *Driving Test*

	s_1: **pass**	s_2: **fail**
o_1: **pay for lesson**	0.8	0.2
o_2: **don't**	0.25	0.75

Given these figures we can apply (2.7) to give the news values for the two options. Here (2.7) takes the form:

$$V(o_i) = Cr(s_1|o_i)V(o_i \cap s_1) + Cr(s_2|o_i)V(o_i \cap s_1). \tag{2.8}$$

So:

$$V(o_1) = (0.8 \times 65) + (0.2 \times 45) = 61$$

$$V(o_2) = (0.25 \times 70) + (0.75 \times 50) = 55$$

$V(o_1) > V(o_2)$: EDT advises you to pay.

We saw before (Section 2.4) that a proposition can be better *news* than one that dominates it over a partition. Similarly, a proposition can be a better *choice* than another that dominates it. It is better news that you don't pay for the lesson and pass than that you pay for the lesson and pass. It is better news that you don't pay and fail than that you pay and fail. But paying is better news, hence a better option, than not paying. The reason is obvious: paying for a lesson increases the probability of passing the exam by enough to make it worthwhile. EDT's advice is natural and correct.

More generally, EDT responds to two things. First, the values of each possible outcome associated with an option. These are the *value* factors $V(o_i \cap s_1)$ and $V(o_i \cap s_2)$ in (2.8). Second, the strength with which that option *promotes* each outcome. These are the *confidence* factors $Cr(s_1|o_i)$ and $Cr(s_2|o_i)$. The merit of an option turns on how it combines them: whether it strongly promotes what you strongly want.

This basic story looks sensible and straightforward. It is attractively austere. But many people think it ignores a metaphysical relation that belongs at the heart of decision theory.

2.7 Newcomb's Problem

The main reason is its performance in this famous case.

> *Newcomb's Problem*. 'Suppose a being in whose power to predict your choices you have enormous confidence. (One might tell a story about a being from another planet, with advanced technology and science, who you know to be friendly, etc.) You know that this being has often correctly predicted your choices in the past (and has never, so far as you know, made an incorrect prediction about your choices), and furthermore you know that this being has often correctly predicted the choices of other people, many of whom are similar to you, in the particular situation to be described below. One might tell a longer story, but all of this leads you to believe that almost

Table 2.5 Values of outcomes in *Newcomb's Problem*

	s_1: **predicted one-boxing**	s_2: **predicted two-boxing**
o_1: **one-box**	99	0
o_2: **two-box**	100	1

Table 2.6 Probabilities of outcomes in *Newcomb's Problem*

	s_1: **predicted one-boxing**	s_2: **predicted two-boxing**
o_1: **one-box**	0.99	0.01
o_2: **two-box**	0.01	0.99

> certainly this being's prediction about your choice in the situation to be discussed will be correct.'[26]
>
> That situation is as follows. There are two boxes. One is opaque. The other is transparent and contains $1,000. You have two options:
>
> o_1: Take only the opaque box ('one-boxing')
>
> o_2: Take the opaque box and the transparent box ('two-boxing')
>
> You get to keep what you take. Yesterday the being predicted what you will now do. If it predicted that you take only the opaque box, then it put $1 million into the opaque box. If it predicted that you take both, then it put nothing into the opaque box. Do you take only the opaque box or both?

We can suppose that the values and probabilities of the outcomes are as in Table 2.5 and Table 2.6.

The probabilities in Table 2.6 are conditional: they indicate your credence (e.g. that the being predicted that you one-box *given* that you one-box). As such, they reflect your 99% confidence that he correctly predicted whatever you do.

Given these figures, EDT recommends one-boxing:

$$V(o_1) = 98.01$$

$$V(o_2) = 1.99$$

This is unsurprising. The being almost certainly predicted correctly. So one-boxing is excellent evidence that you are about to make $1 million, and two-boxing is excellent evidence that you are about to make only $1,000.

[26] Nozick 1970: 207–8.

But obviously one-boxing does nothing to *bring about* any enrichment. Whether or not $1 million is in the opaque box is already settled. Two-boxing won't cause the $1 million to vanish if it is already there; but it *will* cause you to get an extra $1,000. So (you'll say) one-boxing is absurd. More generally, it is absurd to do what merely *indicates* a good outcome irrespective of whether it *brings one about*; so EDT is wrong.

Many philosophers have thought or said such things.

> The 'utility' of an act should be its genuine efficacy in bringing about states of affairs the agent wants, not the degree to which news of the act ought to cheer the agent ... The news of an act may furnish evidence of a state of the world which the act is known not to produce. In that case, though the agent indeed makes the news of his act, he does not make all the news his act bespeaks.[27]

According to them, EDT goes wrong by associating the merit of an act with what it signifies; but it should focus instead on what it brings about. The metaphysical relation it misses is *causality*.

2.8 Causal Decision Theory

The outlines of a remedy (if it *is* a remedy) have been clear since 1972.[28] Causal Decision Theory (CDT) recommends an option not for indicating satisfaction of the agent's wants but for causing it. The following version is just one of various possible realizations.[29] In what follows I'll use 'CDT' to refer to this specific version, and 'the causal approach' or 'the causal theory' for the general idea behind it.

Suppose you are choosing from $\omega = \{o_1, o_2, \ldots o_n\}$ and that you learn all the facts that were causally independent of your choice (i.e. which your choice could not affect). For instance, in *Newcomb's Problem* you would learn irrelevant things like the average temperature in Beijing last night, Queen Victoria's favourite colour before 1861, and so on; but also some relevant things, like what the being predicted, or how much money there is in the opaque box. You would know all that, on this hypothesis. But you would not thereby know anything causally *dependent* on your choice, such as how much money you

[27] Gibbard and Harper 1978: 356–7. For similar claims, see, for example, Skyrms 1984: 63–8, Joyce 1999: 150 (citing the same passage), Weirich 2001: 126–9. For a more balanced treatment of *Newcomb's Problem* than in those works (or in this one) see Elga 2020, which gets across very vividly the perplexity that is probably the proper initial reaction to this case.

[28] Stalnaker 1972.

[29] Lewis 1981a introduces his preferred formulation alongside others that people were discussing in the 1970s. For the most sophisticated recent version of the theory see Joyce 2018. A fuller treatment of Causal Decision Theory would emphasize the 'deliberational' or 'full information' aspect of that theory (see e.g. Joyce 2012); here I set that aside for reasons of space and because it makes no difference to the examples that I'll discuss.

make. Let k^* be the compendious proposition that tells the entire truth about all these causally independent facts.

If you already knew k^*, then any further information you got about your choice would not tell you anything new about what k^* already settles. It would only tell you about things that are causally dependent on your choice. So, if you already knew k^*, the additional value to you of *then* learning an option o would measure the *causal bearing* of o, in your opinion, on what mattered to you. Equivalently, $V(o \cap k^*) - V(k^*)$, which I'll write $U^*_\omega(o)$, measures the causal bearing of o on what you want, given that you know everything that is causally independent of which option you now take.

But you *don't* know everything that is causally independent of your choice. So how can you tell what $U^*_\omega(o)$ is? You can't; but you *can* estimate it. There are many hypotheses specifying how the world could be in all those causally independent ways. If there are finitely many, we can label them $k_1, k_2 \ldots . k_m$. So $K_\omega =_{\text{def.}} \{k_1 \ldots k_m\}$ is a partition of Ω. Exactly one of these is the true k^*, although you don't know which: for each such k_i you have a level of confidence $Cr(k_i)$ that *it* tells the entire truth about everything that you can't affect. You can estimate $U^*_\omega(o)$, the causal bearing of option o on what you want, by taking its expectation i.e.:

$$\mathbb{E}\left(U^*_\omega(o)\right) = \sum\nolimits_{k \in K_\omega} Cr(k)V(o \cap k) - \sum\nolimits_{k \in K_\omega} Cr(k)V(k).$$

Finally, we may simplify by dropping the constant $\sum_{k \in K_\omega} Cr(k)V(k)$.[30] This is because our target decision theory selects options by *comparing* their tendencies to bring about what you want. Addition or subtraction of a constant makes no difference to any comparison. Since in most applications there is a single fixed set of options, we can take ω as read; omitting it from our notation we define:

$$U(o) = \sum\nolimits_{k \in K} Cr(k)V(o \cap k).$$

This expectation $U(o)$ is the **causal utility** of the option o.[31]

Causal Decision Theory advises you to maximize U: choose any option that does most to *bring about* what you want. It tells you to do what is most efficacious, not (like EDT) what is most auspicious.

---

[30] By (2.7), the fact that K is a partition and the fact that $k \cap \Omega = k$ for any $k \in K$, $\sum_{k \in K} Cr(k)V(k) = V(\Omega)$. This additional step would therefore not have been necessary given Jeffrey's conventional setting $V(\Omega) = 0$ (Jeffrey 1983: 99, putting T for Ω). See n. 20.

[31] If K_w is uncountable, then where possible we replace the sums with the corresponding integrals and Cr with the corresponding probability density function.

'Maximize U' isn't very helpful. It might make sense in principle to discuss a partition like K, but any decision theory that depends on your credence in propositions as compendious as the k's is unlikely to be useful in practice.

But we can simplify things. We can 'clump' the k's into broader hypotheses disjoining many k's. More specifically, call any such clumping partition S **suitable for** ω if each cell of S captures everything about the k's that (a) *matters* to you, given your choice, and on which (b) your choice has some *evidential* bearing. Then, if S is suitable for ω, we can show that:

$$U(o) = \sum\nolimits_{s \in S} Cr(s)V(o \cap s) \qquad (2.9)$$

You can calculate the causal utility $U(o)$ more simply using this partition.[32]

[32] More precisely, suppose K_1 and K_2 are partitions such that every cell in K_2 is a union of cells in K_1. And suppose that for each $o \in \omega$ *either* of two conditions holds:

(a) Given any $k_2 \in K_2$ and given o, you don't care which $k_1 \in K_1$ is also true i.e., if $k_1 \subseteq k_2$ then $V(o \cap k_1) = V(o \cap k_2)$.

(b) Given any $k_2 \in K_2$, o is irrelevant to which $k_1 \in K_1$ is also true i.e., if $k_1 \subseteq k_2$ then $Cr(k_1 | k_2) = Cr(k_1 | o \cap k_2)$.

What (a) tells us is that K_2 strips away from K_1 information about factors to which you are indifferent when evaluating o. What (b) tells us is that K_2 strips away from K_1 information about factors to which o is causally (and evidentially) irrelevant. So, a partition of Ω that is suitable for ω is one that can be reached from K by a sequence of clumping operations, each of which has an output partition K_2 that for each $o \in \omega$ stands in relation (a) or relation (b) to its input partition K_1.

What we can show is that for each $o \in \omega$ if (a) or (b) is true then:

(c) $\sum_{k_1 \in K_1} Cr(k_1)V(o \cap k_1) = \sum_{k_2 \in K_2} Cr(k_2)V(o \cap k_2)$

Proof: first suppose (a). Then:

(d) $\sum_{k_1 \in K_1} Cr(k_1)V(o \cap k_1) = \sum_{k_2 \in K_2} \sum_{k_1 \in K_1, k_1 \subseteq k_2} Cr(k_1)V(o \cap k_1)$. But by (a):

(e) $\sum_{k_2 \in K_2} \sum_{k_1 \in K_1, k_1 \subseteq k_2} Cr(k_1)V(o \cap k_1) = \sum_{k_2 \in K_2} \sum_{k_1 \in K_1, k_1 \subseteq k_2} Cr(k_1)V(o \cap k_2)$. And by the probability calculus:

(f) $\sum_{k_2 \in K_2} V(o \cap k_2) \sum_{k_1 \in K_1, k_1 \subseteq k_2} Cr(k_1) = \sum_{k_2 \in K_2} Cr(k_2)V(o \cap k_2)$.

(c) follows from (d)–(f). Now suppose (b). Then again we have (d), and so:

(g) $\sum_{k_1 \in K_1} Cr(k_1)V(o \cap k_1) = \sum_{k_2 \in K_2} Cr(k_2) \sum_{k_1 \in K_1, k_1 \subseteq k_2} \frac{C_r(K_1)}{C_r(K_1)} V(o \cap k_1)$. Now by (2.4):

(h) $\sum_{k_1 \in K_1} Cr(k_1)V(o \cap k_1) = \sum_{k_2 \in K_2} Cr(k_2) \sum_{k_1 \in K_1, k_1 \subseteq k_2} Cr(k_1 | k_2)V(o \cap k_1)$. By (b) and the fact that if $k_1 \subseteq k_2$ then $o \cap k_1 = o \cap k_2 \cap k_1$ it follows that:

(i) $\sum_{k_1 \in K_1} Cr(k_1)V(o \cap k_1) = \sum_{k_2 \in K_2} Cr(k_2) \sum_{k_1 \in K_1, k_1 \subseteq k_2} Cr(k_1 | o \cap k_2)V(o \cap k_2 \cap k_1)$. So by (2.7):

(j) $\sum_{k_1 \in K_1} Cr(k_1)V(o \cap k_1) = \sum_{k_2 \in K_2} Cr(k_2)V(o \cap k_2)$.

For example, in Tables 2.5 and 2.6 the column headings, which say what the being predicted, are plausibly suitable for your options. They specify a state that is causally independent of our choice whether to one-box or to two-box. Of course, other things are *also* causally independent of what you do, other things that you care about: results of football matches and so on, but also, how much money is in the opaque box. But if you know what the being predicted, your choice between one-boxing and two-boxing tells you nothing about *them*. There may also be causally independent questions on which your choice has an evidential bearing, even if you know what the being predicted: for instance, the state of your brain a few seconds ago. But given your choice and what the being has predicted, they too are matters of indifference. So, the clumping partition $\{s_1, s_2\}$ is suitable for $\{o_1, o_2\}$ in *Newcomb's Problem*.

We therefore apply (2.9) to this partition, in conjunction with Table 2.5, to find:

$$U(o_1) = 99Cr(s_1)$$

$$U(o_2) = 100Cr(s_1) + Cr(s_2)$$

Whatever the values of $Cr(s_1)$ and $Cr(s_2)$, the causal utility of two-boxing exceeds that of one-boxing.[33] Causal Decision Theory recommends two-boxing in *Newcomb's Problem*.

The assumption of suitability is not particularly onerous or contentious. In the problems to follow it is obvious that there are, or are meant to be, simple partitions that are suitable for the options in those problems; and I shall generally take for granted what they are.

2.9 Conclusion

That concludes this exposition of EDT. We started with subjective probability and the simplifying idea that you have one ultimate end. We defined the news value of a proposition as the confidence learning that proposition gives you in realizing that end. EDT advises you to maximize news value. But its consequent sensitivity to states that an act *indicates without causing* is widely considered a shortcoming.

This supposed puncture has a preferred patch: the causal approach. This is the orthodoxy. Many people find it obvious that it is rational to choose by considering what your options cause, not what they indicate without causing.

So, by successively applying arbitrarily many clumping operations to K, such that in each case the resulting partition stands to its predecessor in the relation (a) or (b) for each $o \in \omega$, we always end up with a partition S that we can use in conjunction with (2.9) to calculate the causal utility of o for each $o \in \omega$.

[33] By (2.9), $U(o_1) = Cr(s_1)V(o_1 \cap s_1) + Cr(s_2)V(o_1 \cap s_2) = 99Cr(s_1) + 0Cr(s_2) = 99Cr(s_1)$. Similarly $U(o_2) = Cr(s_1)V(o_2 \cap s_1) + Cr(s_2)V(o_2 \cap s_2) = 100Cr(s_1) + Cr(s_2) = 99Cr(s_1) + 1$, since $Cr(s_1) + Cr(s_2) = 1$. So $U(o_2) > U(o_1)$.

To bring out the content of EDT, the next three sections consider applications, each representing a class of cases where (as in *Newcomb's Problem*) it and the orthodoxy diverge. Those discussions have two purposes. Mainly they are meant to give you a vivid picture of what EDT means in practice. But they also occasion arguments that across a range of cases it is EDT and not the causal approach that is getting things right.

3 Self-Signalling

My first class of examples are real or apparent versions of *Newcomb's Problem*: there is a choice between an act that indicates but does not cause a highly desirable state, and one that causes a modest improvement. Sometimes this appearance is deceptive; but in all cases EDT gives defensible advice.

3.1 Medical Newcomb Problems

The first example probably did more than any other to convince people that EDT is obviously wrong.

> *Smoke or Run*. Suppose you like eating eggs, or smoking, or loafing when you might go out and run. You are convinced (contrary to popular belief) that these pleasures will do you no harm at all. (Whether you are right about this is irrelevant.) But also you think you might have some dread medical condition: a lesion of an artery, or nascent cancer, or a weak heart. If you have it, there's nothing you can do about it now and it will probably do you a lot of harm eventually. In its earlier stages, this condition is hard to detect. But you are convinced that it has some tendency, perhaps slight, to cause you to eat eggs, smoke, or loaf. So if you find yourself indulging, that is at least some evidence that you have the condition and are in for big trouble. But is that any reason not to indulge in harmless pleasures?[34]

Suppose you either smoke or run; smoking indicates but doesn't cause an arterial lesion. Notional values and conditional probabilities for outcomes are as in Table 3.1–2: cf. Table 2.5 and 2.6. The situation is analogous to *Newcomb's Problem*, with 'lesion absent' corresponding to \$1 million being in the opaque box and 'run' to

Table 3.1 Values of outcomes in *Smoke or Run*

	s_1: **lesion absent**	s_2: **lesion present**
o_1: **run**	99	0
o_2: **smoke**	100	1

[34] Lewis 1981a: 310–11.

Table 3.2 Probabilities of outcomes in *Smoke or Run*

	s_1: **lesion absent**	s_2: **lesion present**
o_1: **run**	0.6	0.4
o_2: **smoke**	0.4	0.6

one-boxing. And if we hold fixed the presence or absence of the lesion, smoking *makes* you slightly better off than running.

The probabilities are less extreme (and more realistic) than in *Newcomb's Problem*. But the evidential connections are there. Smoking is relatively weak evidence, but still evidence, that the lesion is present; running is evidence that it is not.

So the relevant features of *Newcomb's Problem* are present, and EDT and CDT apparently disagree over what to do. Running has greater news value than smoking because running is evidence that the lesion is absent. But the causal utility of smoking is higher because smoking causes gratification whether the lesion is present or not, whereas running does not.[35]

It is clear what intuition would advise, too: It is silly to avoid smoking just because it is evidence of trouble. There is nothing you can do now about that. Why not indulge in a harmless pleasure? So EDT and common sense look at odds.

But EDT does not give the silly advice. The way the example is described, you know that you will do one of two things:

o_1^*: You run because smoking is evidence of the lesion; or
o_2^*: You smoke because you like it and it does no harm.

Everyone agrees that o_1^* is silly and o_2^* is sensible. But EDT does not prefer o_1^* to o_2^*, because choosing o_2^* over o_1^* is not evidence of the lesion. Any realistic story may allow various ways that a simple physical condition like an arterial lesion may cause smoking. It might cause an increase in the number of acetylcholine receptors. It might even cause a kind of oral fixation. But what it will not do – not unless we are 'got into fairy land' – is cause smoking by controlling your responsiveness to evidential or causal reasons. Nobody ever suggested that in *realistic* versions of *Newcomb's Problem*, a pre-existing health condition might make you sympathetic to CDT.

So the choice between o_1^* and o_2^*, which reflects your responsiveness to those two kinds of reasoning, carries no news about whether you have the lesion.

[35] At least given that $\{s_1, s_2\}$ is a suitable partition for o_1 and o_2, which is as plausible here as in *Newcomb's Problem*. (On 'suitable for' see Section 2.8. From now on I'll assume suitability where necessary without comment.)

Since neither is evidence of anything bad, and since smoking also (you think) does some good, EDT prefers smoking.

Admittedly, the news value of the proposition o_1^-: *that you smoke for no reason at all*, as it were automatically, might exceed the news value of o_2^-: *that you run for no reason at all*. For presumably o_1^- *is* bad news, being evidence of (say) an impulse or urge that plausibly *is* caused by the lesion. But nobody facing *Smoke or Run* is choosing between *those* options. Deliberation is not a process at the end of which you learn that your choice 'just happened'. On the contrary, the upshot of deliberation is a choice whose aetiology you learn along with it. This makes no difference to decision theory in the ordinary run of things: but it does upset Lewis's representation of *Smoke or Run* as a Newcomb-type case.

At least it does, if we doubt that an arterial lesion could influence your choices by controlling your responsiveness to evidential or causal reasons. And with a relatively simple physical condition like that, it *would* be a sensible doubt. But surely there could be other (presumably cortical) derangements that affect choices in this more 'intellectual' way; and it is not much of a stretch to imagine that *they* have undesirable side effects. But then we are back in a genuine Newcomb-type case. Indeed so: but then it is no longer obvious that 'one-boxing' is wrong.[36]

3.2 Continence

This case illustrates what I mean.

> *Beer or Soda.* Today you start work at the Acme Insurance co. After work you and your new boss and colleagues visit the pub. Your colleagues explain that this is a daily fixture that you must attend for as long as you work at Acme. At the pub you can drink beer or soda. Everything else being equal you strongly prefer beer.
>
> But before ordering it, you reflect that you will face an identical decision every weekday night for the foreseeable future; what you choose now therefore strongly signals what you choose in the future. Choosing beer is a strong signal that you will drink often in the future. And you want to avoid that. Still, you know that choosing beer tonight won't *affect* how often you drink in the future. It just *indicates* a weakness for alcohol. You would have that weakness anyway (or not), whether or not you overcome it tonight.
>
> Should you have beer or soda?

Either (s_1) you will be an infrequent drinker or (s_2) you will be a frequent drinker (of beer). You much prefer the first. And either (o_1) you will drink soda

[36] The argument here condenses and reworks that in my 2014a §4.2. In broad outline this line of thought goes back to Price 1991; see also Meek and Glymour 1994: 1007.

Table 3.3 Values of outcomes in *Beer or Soda*

	s_1: **infrequent drinking**	s_2: **frequent drinking**
o_1: **soda**	50	0
o_2: **beer**	60	10

Table 3.4 Probabilities of outcomes in *Beer or Soda*

	s_1: **infrequent drinking**	s_2: **frequent drinking**
o_1: **soda**	0.8	0.2
o_2: **beer**	0.2	0.8

tonight or (o_2) you will drink beer. You like beer more. But drinking beer is a signal that you will be a frequent drinker.

In Tables 3.3 and 3.4 the row headings represent possible present behaviour and the column headings possible future behaviour.

Table 3.3 reflects the fact that holding fixed your future behaviour, you prefer beer. Given infrequent drinking, beer now is better news than soda now (60 vs. 50); the same given frequent future drinking (10 vs. 0). But also: you prefer infrequent future drinking and any present choice (60 or 50) to frequent future drinking and any present choice (10 or 0). Overall Table 3.3 reflects St Augustine's desire that God give him continence, only not yet.[37]

Table 3.4 reflects the fact that present behaviour is diagnostic of future behaviour. Drinking (beer) tonight gives you 80% confidence that you drink often in the future. Abstaining tonight gives you 80% confidence that you drink *in*frequently in the future. These probabilities don't reflect causal influence of present on future behaviour, but rather your opinion that they have a common cause: drinking now is evidence of a taste strong enough to influence you on similar future occasions.

But as at Section 3.1, this is a slightly inaccurate representation of your choice. Given the story I told, the upshot of deliberation will tell you not just whether (o_1) you drink soda or (o_2) you drink beer, but rather one of the following:

o_1^*: You drink soda because drinking beer now is evidence of future drinking; or
o_2^*: You drink beer because you like it, and it does no harm

[37] Ainslie 1991 models this using hyperbolic time preference. For discussion and application of EDT to a generalization of this model see my 2018.

But this time we can ignore this complication. Choosing o_2^* over o_1^* *is* evidence that you drink in the future, because your being responsive (say) to evidential reasoning now is evidence that you will be responsive to it when you face a similar choice in the future, and therefore that you will not drink in the future.

The difference is that in *Smoke or Run* we implausibly took a reflective decision whether to smoke as symptomatic of a physical condition that could only influence *un*reflective decisions. Here we are (more plausibly) taking a reflective decision whether to drink as bearing on future, *equally* reflective decisions on whether to drink. This may be because we think responsiveness to this or that kind of reason is an effect of some complicated but enduring brain state. But it is immaterial whether we can describe that state in any other way. People have been inferring future behaviour from past behaviour since long before anyone thought the brain was involved.[38]

So this *is* a version of *Newcomb's Problem*. In it, EDT recommends the *continent* option (drinking soda) because that is good news; CDT recommends the *incontinent* option (drinking beer) because it 'does no harm'.[39] But now it is hardly obvious that EDT is going wrong. On the contrary, resistance to temptation is a paradigm of common-sense rationality.

One terrible thing about some kinds of weakness – procrastination perhaps, or smoking – is that they can make each individual act of surrender seem rational. Watching television for five more minutes won't affect whether you finish your tax return on time. One more cigarette won't cause you to get a serious illness. These things are nearly always true; and you may *always* have good reason to be confident in them. That can make it seem rational to act as CDT recommends (i.e. to loaf around for five more minutes or to smoke one more cigarette). The trouble is that this reasoning is indefinitely repeatable.

> It does not matter how small the sins are provided that their cumulative effect is to edge the man away from the Light and out into the Nothing. Murder is no better than cards if cards can do the trick. Indeed the safest road to Hell is the

[38] This argument also applies to inferences from your own behaviour to that of causally isolated but sufficiently similar others. So another example of Lewis's, involving *Prisoners' Dilemma, is* a real Newcomb case (Lewis 1979). So too is the decision whether to vote in a large election, at least if voting is inconvenient and has low symbolic value (Quattrone and Tversky 1986; Grafstein 2018). For an interesting argument *against* this interpretation of *Prisoners' Dilemma* and voting cases, see Bermúdez 2018.

[39] It is easy to check that $V(o_1) = 40$, $V(o_2) = 20$. Although we cannot determine numerical values of the causal utilities, still we know $U(o_2) = U(o_1) + 10$.

> gradual one – the gentle slope, soft underfoot, without sudden turnings, without milestones, without signposts.[40]

But according to Evidential Decision Theory, even a tiny step towards Hell is not as rational as it seems, not because it *brings* you appreciably closer to that place but because it *signals* that future steps will. In this real-life *Newcomb's Problem* EDT is giving good advice.

3.3 Managing the News

One natural criticism of this advice is that it involves an irrational policy of managing the news:

> [I]f the prior state obtains, there's nothing you can do about it now. In particular, you cannot make it go away by declining the [beer], thus acting as you would have been more likely to act if the prior state had been absent. All you accomplish is to shield yourself from the bad news. That is useless ... To decline the good lest taking it bring bad news is to play the ostrich.[41]

There are three things that might all be called 'playing the ostrich':

Choosing a because you desire a state s, and a is evidence of s. (3.1)
Choosing a because you desire s and a will cause *belief* in s. (3.2)
Choosing a because you desire to *believe* in s and a will cause belief in s. (3.3)

(3.1) is what EDT recommends in *Beer or Soda*. (3.2) is genuine 'ostrichism'. (3.3) is something that EDT and CDT might both recommend.

Self-signalling as discussed in the economic and psychological literature looks like (3.2) or (3.3). For instance, Bodner and Prelec define self-signalling as 'an action chosen partly to secure good news about one's traits or abilities, even when the action has no causal impact on these traits and abilities.'[42] Bénabou and Tirole attempt to account for altruistic behaviour as arising from 'identity management': a desire to think of yourself as a good person.

> Because people have better, more objective access to the record of their conduct than to the exact mix of motivations driving them, they are led to judge themselves by what they do. When contemplating choices, they then take into account what kind of a person each alternative would 'make them' and the desirability of those self-views – a form of rational cognitive dissonance reduction.[43]

[40] Lewis 1943: 56. [41] Lewis 1981a: 309; see also Pearl 2000: 108–9, Pearl forthcoming 2.
[42] Bodner and Prelec 2003: 105. [43] Bénabou and Tirole 2011: 806–7.

More generally Holton in one place defines self-signalling as 'behaviour that is motivated at least partly by a quest to form beliefs about oneself'.[44]

(3.2) is plainly irrational because it misaligns means and ends on any view. In general, making yourself believe s is neither evidence for nor causally relevant to s itself. But typically EDT does *not* recommend (3.2). EDT may equate the value of a with the conditional probability of the desirable state s, that is, with $Cr(s|a)$. And if EDT recommends a, then realizing a will increase your confidence in s if you update by conditionalizing (Section 2.2). But recommending a for its news value (which EDT does) is not the same thing as recommending a for its effects on your belief. EDT does *not* do this latter thing. On the contrary, EDT might *reject* an option that gives you comforting beliefs.

To illustrate this, and hence the difference between EDT and an ostrich, a fanciful case will do.

> *The Matrix*. You come to suspect that machines are enslaving the human race by keeping their bodies imprisoned inside jars of nutrients, whilst feeding sensory images to their brains that give them the impression they are living in (say) early twenty-first-century America.
>
> You are approached by an elderly gentleman who knows of your suspicions. This character offers you two pills – red and blue. If you take the red pill, he says, you will learn the truth about your situation, whatever it is: it may be that you are indeed a 'brain in a vat', or it may be that things are just as they look – either way, if you take the red pill, you'll find out. But if you take the blue pill then you will simply forget your suspicions and go on to lead a happy though maybe deluded life.

We can represent the values of the outcomes as in Table 3.5. You prefer that we are not enslaved by machines, but either way you want to know. Note that these values for outcomes do not reflect the satisfaction or pleasure or other emotion that you would feel *in* that outcome but rather the values that you now, at the time of decision, attach to that prospect. Even if taking the blue pill would

Table 3.5 Values of outcomes in *The Matrix*

	s_1: **things are as they seem**	s_2: **enslaved by machines**
o_1: **blue pill**	50	0
o_2: **red pill**	60	10

[44] Holton unpublished: 3. Holton 2016 §2 is a more extended philosophical discussion of self-signalling.

Table 3.6 Probabilities of outcomes in *The Matrix*

	s_1: **things are as they seem**	s_2: **enslaved by machines**
o_1: **blue pill**	0.5	0.5
o_2: **red pill**	0.5	0.5

make you happier after taking it – it would alleviate your doubts – you may not now welcome that prospect, because you prefer truth to satisfaction.

Let us turn to probabilities. Which pill you take *is not evidence* for either hypothesis. There is (you think) *no* connection between your choice of pill and the question of whether we are enslaved by machines: neither causes the other, nor is there a common cause. (You don't think, for instance, that any robot overlords would have implanted into humans the desire to take blue pills.) The information that you are about to take the blue pill is therefore evidentially irrelevant to whether the race is enslaved by machines. Table 3.6 reflects this. It follows that

$$V(o_1) = (0.5 \times 50) + (0.5 \times 0) = 25$$

$$V(o_2) = (0.5 \times 60) + (0.5 \times 10) = 35.$$

EDT recommends taking the red pill.

Clearly this is the opposite of ostrichism. 'Managing the news', if it is a basis for criticism, means the kind of self-manipulation involved in taking the *blue* pill. Eschewing the blue pill shows the distance between EDT and any of that.

The same goes for *Beer or Soda*. Contra Lewis, EDT *is not* avoiding beer for the ostrich-like reasons that drinking beer makes you *believe* that you have a weakness for alcohol. It is advising you to avoid beer because drinking it is *evidence for you now* that you have that weakness. As *The Matrix* shows, these are entirely different motivations. It is true that both reasons are *available* in *Beer or Soda*; EDT and ostriches *happen* to agree there. But as we just saw, they disagree elsewhere: what EDT advises is not in general what ostriches do.

3.4 Evidential Blackmail

But though EDT does not recommend the kind of self-manipulation that 'managing the news' typically involves, it does endorse a kind of self-signalling, because it recommends options that you genuinely regard *ex ante* as evidence of states you desire. And even this non–self-deceptive self-signalling seems to invite a subtle kind of exploitation.

> *XOR Blackmail.* You have been alerted to a rumour that your house has a terrible termite infestation, which would cost you $1 million in damages. You do not know whether this rumour is true. A greedy and accurate predictor with a strong reputation for honesty has learned whether it's true, and drafts a letter:
>
>> I know whether or not you have termites, and I have sent you this letter iff exactly one of the following is true: (i) the rumour is false, and you are going to pay me $1,000 upon receiving this letter; or (ii) the rumour is true, and you will not pay me upon receiving this letter.
>
> The predictor then predicts what the agent would do upon receiving the letter, and sends you the letter iff exactly one of (i) or (ii) is true. Thus, the claim made by the letter is true. You receive the letter. Should you pay up?[45]

This looks like a version of *Newcomb's Problem*, where the absence of termites corresponds to the presence of the $1 million and the option of paying up corresponds to one-boxing.

The story is more plausible than that in *Smoke or Run*. Any 'predictor' in *XOR Blackmail* isn't basing his judgment on some simple condition like an arterial lesion: more likely, he is an experienced and perceptive psychologist. It would not be at all surprising if there are people (or algorithms) in existence today that could predict with (say) 55% accuracy how somebody would respond to a letter like this.

Paying up also seems a bad idea. Not only is there no causal connection between paying and not having termites; what connection there is appears to have been constructed by the blackmailer. If you pay up in this situation, it looks as though the predictor can extort from you pretty much at will: he needs only to find some highly undesirable condition *S* that does not obtain (though you don't know this) and then to write (truly) that either *S* obtains or you will pay $1,000.

Finally, EDT seems to recommend paying up. After all, if you are confident that either (i) you have no termites and will pay, or (ii) you have termites and will not, then it looks as though paying up is good evidence that you don't have termites. Although it involves a loss of $1,000, paying up is therefore on balance better news than not, and so is what EDT recommends. So EDT *seems* to recommend something manifestly silly.

Whether it *does* recommend it will depend on your initial beliefs. Suppose you start out – before getting the letter – determined to pay no attention to such things. On receiving the letter you remain confident that you'd never fall for this trick, and so – since you are highly confident that either you have termites or that

[45] Levinstein and Soares 2020: 241 adapting an example from Soares and Fallenstein 2015.

you will pay $1,000 – you also become highly confident that you do have a termite infestation. (Compare: if some reliable forecaster of the weather tells you, in the northern hemisphere in July, that tomorrow it will *either* snow *or* be the hottest day of the year so far, you are likely to believe it.)

In this situation EDT does not recommend paying, because you don't think that paying is evidence that the rumour is false. Paying is evidence that the *letter* is false, though you are in fact highly confident that the letter is true and that you won't pay. So for this reason EDT does not recommend paying, even if you get the letter and even if you believe it. And there will be recipients of the letter who are confident that the letter is true, confident that they are not going to fall for it, confident that they are following the advice of EDT, and right on all three counts.[46] Of course, if you start out with this belief and your house is termite-free, you would never have got the letter in the first place: that, presumably, is the position that most sensible followers of EDT are actually in.

EDT also gives this sane advice to people whose starting point is opinionated in the other direction. Suppose you are confident you *would* fall for some such scam. Getting the letter makes you highly confident that you will pay up but also that your house is termite-free. Again, therefore, you believe the letter, this time because you believe (ii). But not paying isn't evidence that the rumour is true – it is (again) evidence that the letter is false. Of course, if the letter is true, then you will pay; but that is hardly grounds to criticize EDT, which plainly advises you not to.[47]

In short, EDT only advises paying if getting the letter makes your confidence in the disjunction that it asserts not just high, but high conditional on both options: you must find (i) or (ii) likely given that you pay up, and likely given that you don't. Nothing in the letter as described, or in the back story as told, gives us reason to expect that. So sending the letter to anyone of whom it is true is not an easy way – it is not any way – to extort people who follow the advice of EDT: many followers of that theory will just not get the letter; and of those who do, many will not pay because it convinces them that they have termites.

[46] For instance, let p be the proposition that you pay up and t the proposition that you have termites. Then receiving the letter may give you the following credences: $Cr(p \cap t) = 0.09$, $Cr(p \cap \bar{t}) = 0.01$, $Cr(\bar{p} \cap t) = 0.81$, $Cr(\bar{p} \cap \bar{t}) = 0.09$. In this situation you are confident that what the letter says is true ($Cr((p \cap \bar{t}) \cup (\bar{p} \cap t)) = 0.82$), but you don't think that paying up is *evidence* that you don't have termites ($Cr(t|p) = Cr(t|\bar{p}) = 0.9$), so EDT doesn't advise it.

[47] For an example (notation as in n. 46): suppose that on receiving the letter your credences become, $Cr(p \cap \bar{t}) = 0.81$, $Cr(\bar{p} \cap t) = 0.01$, $Cr(\bar{p} \cap \bar{t}) = 0.09$. In this situation you are confident that what the letter says is true ($Cr((p \cap \bar{t}) \cup (\bar{p} \cap t)) = 0.82$), but you don't think that paying up is evidence that you do not have termites ($Cr(t|p) = Cr(t|\bar{p}) = 0.1$), so EDT doesn't advise it.

Still, I haven't said anything about how one gets into that position in the first place: that is, how one gets into the position either of being highly confident *ex ante* that you will pay, or of being highly confident *ex ante* that you won't. And the trouble is that EDT by itself is no guarantee of it: for there are possible recipients of the letter whom EDT *does* advise to pay. Suppose that on getting the letter you become about equally, and quite highly, confident of each of (i) and (ii) – say, 45% confident that (i) is true and 45% confident that (ii) is true.[48] Then you *will* consider paying to be evidence that there is no termite problem.

Here then, we may have an example of *Newcomb's Problem* where EDT is giving bad advice. But how bad is it, really? Levinstein and Soares write that '[h]ow frequently people end up with termite infestations does not change regardless of their willingness to pay blackmailers of this kind'. True, but how frequently people *who get this letter* have termite infestations does indeed co-vary with their willingness to pay (at least it does if these statistics reflect the credences that we are now imagining).[49] Amongst people who get the letter, those who pay do not 'end up' with termites, whereas those who do not pay typically do.

But still, paying seems worse than one-boxing in the Newcomb case for the following reason: in the Newcomb case, there is at least some kind of causal connection between your choice and what is in the opaque box; namely, that they have a common cause. Whereas in *XOR Blackmail* the only connection between whether you send the money and whether you have termites looks like one that the blackmailer has simply made up. No responsible decision theory should be sensitive to connections that are so easy to fabricate.

But what makes the connection is that you have received the letter; and given his own constraints this is *not* something that the blackmailer can make true at will. He only writes to people of whom either (i) or (ii) is true; it is not up to him who (if anyone) belongs in either class. So whether you receive the letter is causally sensitive to (a) whether your house is termite-free; (b) whether you are disposed to pay up on getting it: you get the letter if (a) and (b) are both true or both false.

Here is the (zigzag) causal connection between the presence of termites and your choice whether to pay. See the arrows marked in bold in Figure 3.1. There is a signal (the letter) that is a common *effect* of the former and of a psychological state that causes the latter: (a) and (b) are uncorrelated in the overall

[48] Of course, you cannot be more than 50% confident in each of (i) and (ii), because you know for certain that they are not *both* true.

[49] I mean: we consider a population in which the relative frequencies of termite infestations, willingness to pay,and so on reflect your credences that they apply to you. For more on what it means for 'statistics to reflect credences' see Greene 2018.

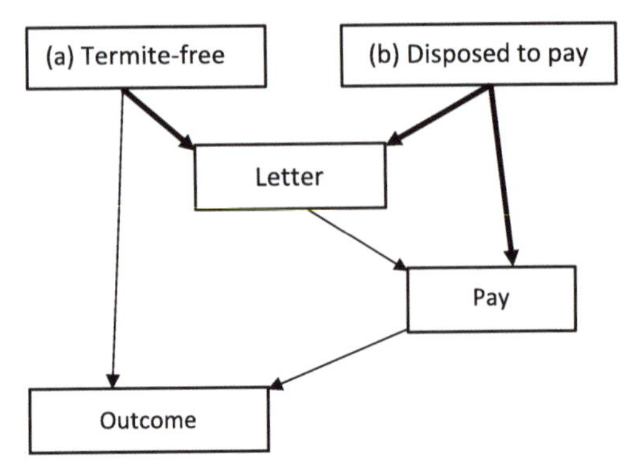

Figure 3.1 Causal dependencies in *XOR Blackmail*

population but correlated given the presence (or given the absence) of this signal. The unfamiliarity of this arrangement may be one reason why EDT's advice seems strange.[50]

It may seem less strange when we compare an admittedly fantastical case with a relevantly similar causal structure.

> *Cassandra.* She is a reliable clairvoyant that you invited to your party. Usually she is (as one would imagine) a most entertaining guest, but tonight she seems anxious. You ask her what's wrong. 'Somebody at this party is going to have a serious accident tonight', she whispers. 'I couldn't make out who. But I did see that they were wearing a limited-edition Batman-themed watch.' To your horror you realize that *you* are wearing such a watch!
>
> Later that evening you find yourself conversing with your colleague Jones, also at the party. You and Jones have been bitter rivals ever since his team beat yours at the works pub quiz in Swindon ten years ago. You know (and take pleasure in the fact) that Jones is particularly covetous of your Batman watch. Suddenly it occurs to you that you could offer it to him. The offer would astonish him, but he would certainly accept it then and there, and he would wear it smugly

[50] Of course, it is true that your choice whether to pay, and the *outcome* of your choice, have a common element in their causal history, namely the absence or presence of termites (see Figure 3.1). In this respect the situation *does* resemble *Newcomb's Problem*. What distinguishes it from *Newcomb's Problem* is that in *XOR Blackmail*, *this*, more direct, causal connection between termites and payment is decision-theoretically irrelevant, since your knowledge of whether you have received the letter 'screens off' any evidential bearing that your choice has on the presence of termites via that channel. The only causal path between termites and your choice that matters to EDT is the zigzag one highlighted in Figure 3.1. An analogous version of *Newcomb's Problem* would be a case involving *two* predictors: one, whose predictive abilities are no better than chance, decides what money goes into the opaque box; and a second, whose predictive abilities are near-perfect, tells you whether the first predictor got it right. XOR Blackmail corresponds to the case where the second predictor tells you, before you choose, that the first predictor got it right. (I am grateful to a referee for helping me to think more clearly about this.)

for the rest of the evening – or at least (you grimly reflect) until he got what was coming to him. Should you offer Jones the watch?

You know there is no direct causal connection between anyone's wearing the watch and their suffering an accident. Nor could these events have a common cause. But Cassandra's vision is their common effect; and learning what you just learnt about it makes wearing the watch evidence of a forthcoming accident. So although giving the watch to Jones neither causes a serious accident to befall him nor prevents one from befalling you, EDT still says that you should do it.[51]

There is some plausibility in this advice: after all, people in situations like this who take it (and only they) will evade the accident that befalls someone in their party later that evening. But the case is structurally like *XOR Blackmail*. Here too you learn of some common effect of two uncorrelated events; and the effect is to make it rational to act on the correlation between them that this learning episode creates. So if the advice makes sense in one case, it should make sense in the other.

Still, it is worth repeating that EDT only advises paying in *XOR Blackmail* to subjects who start out uncertain how they would respond to the letter. If you are resolute from the outset that you won't pay up, then even if you get it and even if you believe it, EDT advises you not to pay. Anyone whose intuition about this case is firmly against paying will therefore find that if she ever has the misfortune to face it, EDT will certainly not give her bad advice.

4 Randomization

EDT takes full account of the likely causes as well as the likely effects of your current decision. Whereas the causal orthodoxy treats the current decision as though it were completely isolated from external influence – as if it had effects but no causes. We have seen one consequence of this: EDT allows for, and CDT ignores, any expected covariation between your choice and any state of the world that you cannot affect.

Another consequence, which this section explores, is that the causal orthodoxy makes it hard to see any point in *breaking* the connection between a

[51] This case involves backwards causation, from the later occurrence of an accident to the earlier occurrence of the vision. It also involves a causal cycle, running from (a) Cassandra's vision of the watch, to (b) your decision of how to dispose of the watch, to (c) the wearing of the watch by whoever happens to suffer the accident, and then from (c) back to (a). Cases involving backwards causation and causal cycles are at best highly speculative; so it would be reasonable to dismiss the example *if* it were being presented as a serious objection to the causal theory. However, *this* case is intended only to illustrate that it can be intuitive to respond in the way that EDT does to a causal structure that is relevantly like that in *XOR Blackmail*. To that end, a fantastic case like will do, *if* it elicits clear intuitions, as I hope *Cassandra* does. (I thank a referee for helpful comments.)

current decision and its influences. From a perspective that pretends that there *are* no such influences, trying to break them can only look like wasted effort. But EDT *can* make sense of the efforts people make in that direction, as the next two examples illustrate.

4.1 Divination

Here is a stylized version of a problem some people may have faced.[52]

> *Hide and seek.* You are tracking a herd of bison. The trail is cold, but you know that the herd went north or south – it is 50–50 which.
>
> The bison around here (you also know) can anticipate your choices. The predictive mechanism is unknown, but you think they have evolved sensitivity to local conditions that also bias human choices. For instance, very subtle – not consciously detectable – erosion patterns in local rock formations might influence you to go north; if so, they also (again unconsciously) influence the bison to go south.
>
> You have two obvious options: go north directly or go south directly. But there is a third option: you can, at a small expense of time, consult a randomizing device: an 'augur' who is specially authorized to count the birds on the nearest tree. If the number of birds is even, you go north. If odd, you go south.

What matters to you is (i) whether you catch the bison – that matters very much – and (ii) whether the augur uses up time to count birds. See Tables 4.1 and 4.2.

The column headings specify whether the bison go north or south / whether the number of birds on the nearest tree is even or odd. As usual, each entry in Table 4.1 indicates the news value of the proposition combining its row and column headings. Each entry in Table 4.2 indicates your confidence in its row heading given the column heading.

In the best outcomes you catch the bison without augury. For instance, in $o_1 \cap s_2$, the bison go north, and you go directly north (so catch them): this scores 80. In the second-best outcomes, the augur gives good advice: for instance (as per $o_3 \cap s_4$), the augur counts an odd number of birds and the bison go south. That scores 70: 80 for catching the bison, minus 10 for loss of time. Similarly, in the worst and second-worst outcomes you lose the herd respectively with and without the additional cost of augury.

The numbers in Table 4.2 reflect the correlation between where the bison go and where, if anywhere, you go directly (without augury). For instance, the entry corresponding to o_1 and s_2 says that $Cr(s_2|o_1) = 0.1$: your confidence that the bison go north and there is an odd number of birds, *given that* you go directly

[52] This is a more realistic version of my 2014b.

Table 4.1 Values of outcomes in *Hide and Seek*

	s_1: **N / even**	s_2: **N / odd**	s_3: **S / even**	s_4: **S / odd**
o_1: **go directly N**	80	80	10	10
o_2: **go directly S**	10	10	80	80
o_3: **randomize**	70	0	0	70

Table 4.2 Probabilities of outcomes in *Hide and Seek*

	s_1: **N / even**	s_2: **N / odd**	s_3: **S / even**	s_4: **S / odd**
o_1: **go directly N**	0.1	0.1	0.4	0.4
o_2: **go directly S**	0.4	0.4	0.1	0.1
o_3: **randomize**	0.25	0.25	0.25	0.25

north. It is low because although you don't know what the local erosion patterns *are*, you know that a direct choice of direction would be subconsciously responsive to them, and that bison are also sensitive in the opposite way: if you go directly north, they go south, and so on. So, you have only (say) 0.2 confidence that if you go directly north you will catch them. And you and the augur know perfectly well that it is a completely independent question, on which neither your choice nor the bison's has any bearing, whether there is an odd or even number of birds in the nearest tree. Since you are 50–50 on *that*, it follows that your confidence in s_2 given o_1 is 0.1.

The last row of Table 4.2 covers what could happen if you randomize. Doing so breaks the stochastic connection between you and the bison. If you go north at the augur's behest, the bison are as likely to go north as to go south; ditto if you go south. Since the augur himself is equally likely to advise north or south, all four possibilities are equally likely.

What does EDT recommend? We have

$$V(o_1) = V(o_2) = (0.1 \times 80) + (0.1 \times 80) + (0.4 \times 10) + (0.4 \times 10) = 24$$

$$V(o_3) = (0.25 \times 70) + (0.25 \times 0) + (0.25 \times 0) + (0.25 \times 70) = 35.$$

In fact, it is obvious that if you are 50–50 as to whether the augur advises north or south, then *whatever* the credences in the bottom row of Table 4.2, $V(o_3) = 35$.[53] So given the first two rows of Table 4.2, EDT recommends using the augur in all such cases.

[53] Proof: write $x_i = Cr(s_i|o_3)$. The condition means: $x_1 = x_2$ and $x_3 = x_4$. It follows that $x_1 + x_4 = x_2 + x_3 = 0.5$. Therefore $V(o_3) = 35$.

This is unsurprising: going directly north (or south) is bad news because it is a strong signal that the bison went the other way. Consulting the augur is better news: by breaking the stochastic connection between where you and the bison end up, it signals that you have entered a fair lottery rather than one that is rigged against you.[54]

But the causal approach recommends against augury. Clearly you cannot now affect which of s_1–s_4 is true: the bison have already gone; the birds are there in the trees. So if you are equally confident in each of s_1–s_4 then:

$$U(o_1) = U(o_2) = 45$$

$$U(o_3) = 35.$$

If (like any sensible person) you think there is no correlation between where the bison are and what the augur says, then *whatever* your credences in s_1–s_4, randomizing *never* maximizes causal utility: CDT always recommends going directly north or going directly south.[55]

This too is unsurprising. The causal theory behaves as though your choice is isolated from external influences. This includes the subconscious influences that the bison subconsciously exploit. From the causal perspective, paying even a small cost to disrupt the resulting correlations must seem gratuitous. EDT therefore does, and the causal theory does not, rationalize randomization in cases like *Hide and Seek*. But if such cases do arise, randomization *is* plausible: but only Evidential Decision Theory appreciates this.

Could such cases arise? It wouldn't take much. There must be a material state of nature that is uncertain. The subject's behaviour must be correlated with the state of nature, and not because the behaviour *causes* the state but because the state responds to whatever causes some costly-to-eliminate bias in the subject. The correlation must tend to make things worse. And there must be an option to break the correlation. That's all it would take.

---

[54] As with *Beer or Soda*, no difficulty arises from the fact that by choosing one learns something about one's responsiveness to reasons. When I learn, by going directly north, that going directly north is the upshot of my deliberation, I may also learn the reason (whatever it was) that decided me not to randomize. But knowing my reason for *not randomizing* does nothing to sever the probabilistic connection between the bison's having gone south and my *not randomizing in this particular way* (i.e., by going directly north). So, the grounds for denying that *Smoke or Run* is a Newcomb-type case don't endanger my claims about what EDT recommends in *Hide and Seek*.

[55] Proof: write $p_i = Cr(s_i)$. Then 'no correlation' means $p_1/(p_1+p_3) = p_2/(p_2+p_4)$, hence (a) $p_1p_4 = p_2p_3$. Now $U(o_3) \geq U(o_1)$ implies $70p_1 + 70p_4 \geq 80p_1 + 80p_2 + 10p_3 + 10p_4 = 70p_1 + 70p_2 + 10$. Therefore (b) $p_4 > p_2$. Similarly, $U(o_3) \geq U(o_2)$ implies (c) $p_1 > p_3$. But no negative quantities occur in (b) or (c); so $p_1p_4 > p_2p_3$ contra (a). Therefore $U(o_1) > U(o_3)$ or $U(o_2) > U(o_3)$.

Some anthropologists suggest that such cases *do* arise, and that when they do, they may explain (and not only justify) actual divinatory practices. O. K. Moore first applied this idea to the hunting practices of the Naskapi; more recently Dove has done the same for the Kantu' of Indonesian Borneo.[56] A similar case might arise if you are playing 'Rock, paper scissors' against an expert, or a sufficiently powerful computer.[57] Or again, it may even be rational for a US President, fearing that foreign agents are among his policy advisers, to leave important decisions to his (or Nancy's) personal astrologer.

But even if you are neither a US President nor a reader of entrails, randomization will matter in other ways that the causal approach ignores. Many people on appointments committees know about unconscious bias (e.g. towards candidates who share your class, gender, religion). Replacing later stages of an appointment process – or university admissions – with a suitably weighted lottery could cheaply eliminate that kind of unfairness. Again, there is plenty of evidence that our seemingly random choices between seemingly indifferent options are accurately predictable using simple heuristics. It would be naïve to doubt that corporations and states are already using these devices to extract our attention, time and money.[58] Randomization could preserve or enhance autonomy in these situations. Finally, and as I now discuss, randomization plays a well-established epistemological role in the discovery of *causal* relations. There is an irony in the fact that only *Evidential* Decision Theory vindicates this procedure.

4.2 CDT v. RCT

The following case illustrates the point.

> *The Great Lorenzo*. He claims to be *lacto-pathic*: from one sip of a cup of tea with milk he can tell whether milk or tea was poured first. You hypothesize that the order of pouring does indeed have this effect on The Great Lorenzo.
>
> To test this, you make tea in three identical-looking teacups, two tea-first and one milk-first. The Great Lorenzo enters and, after one sip from each, must identify the milk-first cup. If he is lacto-pathic, he will certainly be able to do this. If he is not lacto-pathic, he will have to guess.

[56] Moore 1957; Dove 1993. E.g., Moore writes: 'If it may be assumed that there is some interplay between the animals they seek and the hunts they undertake, such that the hunted and the hunters act and react to the other's actions and potential actions, then there may be a marked advantage in avoiding a fixed pattern in hunting. Unwitting regularities in behaviour provide a basis for anticipatory responses. For instance, animals that are 'overhunted' are likely to become sensitized to human beings and hence quick to take evasive action. Because the [outcome of divination] and the distribution of game are in all likelihood independent events, i.e., the former is unrelated to the outcome of past hunts, it would seem that a certain amount of irregularity would be introduced into the Naskapi hunting pattern by this mechanism' (Moore 1957: 71).

[57] Wang et al. 2014. [58] Poundstone 2014 is an entertaining discussion.

> You must choose which cup gets the milk-first 'treatment'. You have four options: either directly choose one of the three cups, or use a randomizing device – say, a milk-pouring robot.
>
> It is unlikely but possible that some bias will affect both your choice (if you choose a cup) and Lorenzo's (if he guesses one). That is: a direct choice of cup is subconsciously caused by some peculiarity of that cup – a tiny chip in the rim, or slightly brighter colouring. And if Lorenzo isn't lacto-pathic, *he* might notice and consciously choose the peculiar cup. Or he might not notice, and 'honestly' guess, but his guess is biased like yours, so he still passes. But whether it affects The Great Lorenzo subconsciously or consciously, the point is that it also affects *you* subconsciously, and so increases the probability of a spurious correlation between your choice and his.
>
> What matters is just that the test deliver a true result: either Lorenzo is lacto-pathic and passes, or he is not and he fails. How should you allocate the treatment?

There are four options: one for each teacup plus one for randomization. Randomization itself divides into three possibilities, depending on which cup the robot selects.

There are four possible states, each of which, together with your option, settles what matters. If Lorenzo is lacto-pathic, then he will choose the right cup, whichever one it is and whether or not the robot was involved. So if he is lacto-pathic, you will get a true positive. If not, then whether the outcome is a false positive or a true negative depends on which cup he chooses.

Label each state y or (n, i):

y: he is lacto-pathic

(n, i): he is not lacto-pathic and chooses cup i

(The cups get arbitrary numbers, not visible on the cups themselves.) The outcome values are as in Table 4.3.

In Table 4.3 the row headings o_1, o_2, o_3 correspond to the options: directly pouring milk first into cup 1, into cup 2, and so on. r corresponds to

Table 4.3 Values of outcomes in *The Great Lorenzo*

		y	$(n, 1)$	$(n, 2)$	$(n, 3)$
o_1		1	0	1	1
o_2		1	1	0	1
o_3		1	1	1	0
r	r_1	1	0	1	1
	r_2	1	1	0	1
	r_3	1	1	1	0

Table 4.4 The Great Lorenzo: probabilities of outcomes

		y	$(n, 1)$	$(n, 2)$	$(n, 3)$
o_1		0.4	0.24	0.18	0.18
o_2		0.4	0.18	0.24	0.18
o_3		0.4	0.18	0.18	0.24
r	r_1	0.4	0.2	0.2	0.2
	r_2	0.4	0.2	0.2	0.2
	r_3	0.4	0.2	0.2	0.2

randomizing. r allows three possibilities for which cup gets 'milk before tea' – the subsidiary headings r_1, r_2, r_3. For instance, r_1 corresponds to the possibility that you randomize and the robot chooses cup 1.

The values reflect your desire for a *true* positive or *true* negative result: either Lorenzo *is* lacto-pathic and *passes*, or he *is not* lacto-pathic and *fails*. Your value for such outcomes is 1; for the rest 0. For instance, in the outcome corresponding to o_1 and y, you choose to pour 'milk first' into cup 1 and Lorenzo is lacto-pathic, so he correctly identifies cup 1: a true positive. Similarly, in the outcome corresponding to o_1 and $(n, 1)$, you choose cup 1 and Lorenzo correctly guesses that cup (i.e. a false positive, which scores zero).

These values are simplifications, but they are plausible enough. For instance, suppose you must bet your last dollar for or against the proposition that The Great Lorenzo is lacto-pathic, and how you bet must depend only on whether Lorenzo passes this test.[59]

The probabilities of outcomes conditional on options (or on their sub-possibilities) are in Table 4.4. I suppose you start out 40% confident that Lorenzo is genuine.

You suspect that if you directly choose a cup, then if Lorenzo is not lacto-pathic, subliminal cues *may* direct him to it. That is why the last three entries in the o_1 row are 24%, 18%, 18%. Given that you choose cup 1, you are 60% confident Lorenzo is not genuine. But within this 60% you give slightly more credence to the possibility that he gets it right than to either possible way he could err.

[59] Still, the argument of this section will *not* apply when the value of an outcome reflects not an all-or-nothing concern that the experiment deliver a true result but a more fine-grained concern that it improves the accuracy of the subject's *credences*. In this slightly more complex case CDT *does* value randomization; but the problem is that it overvalues it: we can construct other cases where, from the point of view of accuracy, it makes obvious sense *not* to randomize, and yet CDT still advises it. For details of this argument see Appendix B.

Randomizing eliminates this distortion. It breaks any correlation between the true milk-first cup and Lorenzo's choice arising from a subliminal common cause. If he isn't lacto-pathic then his guess against a randomizer is no better than chance. Whichever cup the robot nominates, Lorenzo (if fraudulent) is equally likely to choose any. So your 60% confidence that he is not lacto-pathic divides equally between these possibilities: see the last three entries in the r_1, r_2 and r_3 rows.[60]

We calculate the news values of your options using (2.7). Your options are: (o_1) choose cup 1 directly, (o_2) choose cup 2 directly, (o_3) choose cup 3 directly and (r) randomize.

$$V(o_1) = V(o_2) = V(o_3) = 0.76$$

$$V(r) = 0.8$$

EDT recommends randomizing. This is not surprising. Choosing directly is evidence of subliminal influences that may similarly influence the Great Lorenzo; evidence therefore of a false positive.[61]

What about the causal utilities? Notice that which column heading obtains is causally independent of your choice. For suppose Lorenzo is lacto-pathic. Then he would have been lacto-pathic whatever you had done. (How you pour tea won't affect his powers!) Or suppose he is *not* lacto-pathic and nominates cup 1. Then his guess was either random or responsive to some pre-existing feature, such as a tiny flaw in the cup. Either way, it is not something that your choice could affect. So we may apply (2.9) to Table 4.3.

Doing so seems to require knowing your unconditional credence in each state. It doesn't: *however* confident you are in each of y, $(n, 1)$, $(n, 2)$, $(n, 3)$, at

[60] Kadane and Seidenfeld seem to reject this point, as follows. Suppose, for example, you expect Lorenzo (if a fraud) to choose cup no. 1 because of a difference in colouring. Then putting the milk first into cup no. 1 should mean that he is (if a fraud) more likely than chance to choose no. 1, *whether that allocation was deliberate or random* (1990: 335–6). – True, but irrelevant to the argument for Table 4.4, which assumes that you are completely unsure *which* cup Lorenzo is biased towards (if any). All you know is that there is a good chance it is the same as whichever one *you* choose. So given that the *robot* chooses this or that cup, you *remain* ignorant as to which cup Lorenzo is biased towards. This is consistent with the fact that for each i, pouring milk first into cup i at random is no better than doing so directly, *given* that Lorenzo is biased towards cup i. But it shows that the situation has a Simpson-paradoxical structure. For let b_i be the proposition that Lorenzo is a fraud and biased towards cup i. Then for each i, $Cr((n,1)|b_i \cap o_1) = Cr((n,1)|b_i \cap r_1)$, whereas $Cr((n,1)|o_1) > Cr((n,1)|r_1)$.

[61] Details of the calculation: $V(o_1) = V(Y \cap o_1) + \sum_{i=1}^{3} V((N,i) \cap o_1)Cr((N,i)|o_1) = 0.4 + 0.18 + 0.18 = 0.76$, and similarly for o_2 and o_3. Since $r_i \cap r = r_i$ for each $i = 1, 2, 3$ it follows from (2.7) that $V(r) = \sum_{i=1}^{3} V(r_i)Cr(r_i|r)$. Applying (2.7) again, $V(r_1) = (y \cap r_1) + \sum_{i=1}^{3} V((n,i) \cap r_1)Cr((n,i)|r_1) = 0.4 + 0.2 + 0.2 = 0.8$; similarly for r_2 and r_3. So $V(r) = \sum_{i=1}^{3} 0.8Cr(r_i|r) = 0.8$.

least one 'direct' option has at least as much causal utility as randomization. For let c_1, c_2 and c_3 be your credences in $(n, 1)$, $(n, 2)$ and $(n, 3)$. Then:

$$U(o_1) = Cr(y) + c_2 + c_3$$

$$U(o_2) = Cr(y) + c_1 + c_3$$

$$U(o_3) = Cr(y) + c_1 + c_2.$$

This is all intuitive. Suppose you directly choose cup 1. Then, if Lorenzo is lacto-pathic, you get a true positive. If not, you get a true negative if and only if he chooses either cup 2 or cup 3. So the causal utility of choosing cup 1 is the sum of your credences in these three ways your experiment could tell the truth.

Next, note:

$$U(r) = Cr(y) + \frac{2}{3}c_1 + \frac{2}{3}c_2 + \frac{2}{3}c_3.$$

Again, this is intuitively plausible. Suppose you randomize. If Lorenzo is lacto-pathic, then you are guaranteed a true positive. If he is not lacto-pathic and chooses cup 1, you get a true negative if and only if the robot poured milk first into cup 2 or into cup 3; there is a one-third probability of each of these. Similarly, if he is not lacto-pathic and chooses cup 2 or cup 3, there are again two ways to get a true negative, each with probability one-third. So the causal utility of randomizing is the sum of your credences in each of these seven ways that you could get a true result.[62]

It follows from these equations that $U(r)$ cannot exceed $U(o_1)$, $U(o_2)$ *and* $U(o_3)$.[63] Always at least one 'direct' option has as much causal utility as randomizing. And this is intuitive. The *effects* of picking out this or that cup for treatment are the same whether it is you who chooses or the robot that chooses. Since the causal approach cares *only* about the effects of your choices, naturally it sees no point to randomizing.

This should trouble followers of the causal approach, who seem forced to reject a perfectly standard experimental protocol for which there are obvious and compelling grounds. Admittedly they need not reject *all* such grounds; there may in other circumstances be other reasons to randomize the

[62] The first three equations are obvious consequences of (2.9). Here is a formal argument for the fourth.

(a) $V((n,i) \cap r) = \sum_{j=1}^{3} Cr(r_j | (n,i) \cap r) V((n,i) \cap r_j) = \frac{2}{3}$ by (2.7) and Table 4.4.

(b) $U(r) = Cr(y)V(y \cap r) + \sum_{i=1}^{3} V((n,i) \cap r) Cr((n,i))$ by (2.9).

(c) $U(r) = Cr(y) + \sum_{i=1}^{3} \frac{2}{3} c_i$, by (a), (b) and Table 4.3.

[63] Proof: given the U values for the individual options we have $3U(r) = U(o_1) + U(o_2) + U(o_3)$, but this is impossible if $U(r) > U(o_1), U(o_2), U(o_3)$.

administration of an experimental treatment among test subjects.[64] But even if one's only concern is to get at the truth about what is causing what, it *ought* to matter that

> randomization changes the possible causal interactions in an experimental situation. Unrecorded, subliminal cues cannot influence the treatment allocation. One type of experimenter bias becomes impossible.[65]

EDT is wholly receptive to this point. The causal theory is not. Experiments conducted under its auspices are more probably misleading.

It is worth briefly mentioning one line of response, which is that the causal approach *is* receptive to this point 'from a more distant perspective'. Suppose that we are *planning* the experiment that is supposed to test The Great Lorenzo. At this planning stage, we must choose between (a) letting the experimentalist *choose* which cup gets the treatment, and (b) forcing the experimentalist to *randomize*. From this perspective, Causal Decision Theory prefers to randomize! Intuitively, this is because the *effect* of letting the experimentalist choose is to make a false positive more likely than if we were to entrust everything to the robot.[66] So, in practice, CDT may create no difficulty because it recommends that we *plan* to randomize in our future experiments, even if it recommends that we abandon this policy when we get there.

Two points in reply. First, it is easy to imagine situations in which there is a cost involved in binding a *future* experimentalist (that may be one's own future self) to accord with one's *current* plans. (For instance, suppose that we must pay this future person to follow our instructions rather than to act on his own initiative.) CDT would then incur this cost, whereas EDT would not need to: it could get the same outcome for free. Whether or not this difference between EDT and CDT is best described as a difference between 'rational' and 'irrational' recommendations, it still puts EDT at a clear advantage.

---

[64] For instance, it may reduce costs to consumers of the experiment (Kadane and Seidenfeld 1990) or be a rational response to memory limitations (Icard 2021). Fisher 1935 is the classic defence of randomization on grounds that include those emphasized here.

[65] Swijtink 1982: 163.

[66] More formally: write a for the proposition that we leave the choice of treatment to the experimentalist, and write b for the proposition that we force him to use the robot. Write $a > o_i$ for the proposition that if we were to leave it to the experimentalist, then he would assign the treatment to cup i; write $b > r_i$ for the proposition that if we were to force him to use the robot, then the robot would assign the treatment to cup i. Then we have $U(a) = Cr(y) + \sum_{i \neq j} Cr((n,j) \cap (a > o_i))$ and $U(b) = Cr(y) + \sum_{i \neq j} Cr((n,j) \cap (b > o_i))$. But if $i \neq j$, then clearly $Cr((n,j) \cap (b > o_i)) > Cr((n,j) \cap (a > o_i))$: since there is *some* chance of a common cause (a flaw in one of the cups) prompting both Lorenzo and the experimentalist in the same direction, we can expect that $Cr((n,i) \cap (a > o_i)) > Cr((n,i) \cap (b > o_i))$, whence it follows that $Cr((n,j) \cap (b > o_i)) > Cr((n,j) \cap (a > o_i))$ if $i \neq j$. Therefore $U(b) > U(a)$.

Second, there is in any case something troubling about a decision theory that treats the same problem (whether to randomize) in these totally opposing ways, depending on whether we consider it *ex ante* or *in media res*. Surely, whether you are *considering* the same decision problem in advance or as a present problem cannot make a difference to what you ought to do. That it does, is a troubling consequence of CDT; and in Section 5 I say in more detail both why it is troubling and why it is a consequence.

Before that, let me turn to some more general concerns about randomization itself. We saw two types of case where EDT prefers to randomize; and it seemed clear that in both cases EDT is getting it right. But is it? There are two strong arguments *against* randomization; and although not widely accepted in practice, they *are* persistently defended in theory, and it is worth considering briefly whether they should trouble EDT.

4.3 Against Randomization

The arguments are as follows: (a) the Bayesian argument that mixing optimal with suboptimal options can never itself be optimal; (b) a pragmatic argument that what randomization buys can be got without it. I'll reject (a) and endorse (b); but (b) does nothing to harm EDT – or to help the causal approach, which also rejects these other methods.

(a) Bayesian Argument. Informal version: suppose you can randomize between two options *h* and *t*, say by tossing a coin: heads you get *h*, tails you get *t*. Now either one of *h* and *t* is better than the other (by whatever standard you are using) or they are equally good. If one is better, then performing it directly must be better than probabilistically 'mixing' it with the inferior option. And if *h* and *t* are equally good, then randomizing may be no *worse* than either; but it also is no *better*.[67]

For example, in *Hide and Seek* either you're more confident that the bison have gone North, or you're more confident that they have gone South, or you're 50–50. In the first case, how could paying to randomize be better than going North directly, for free? In the second case, how could it be better than going South directly? In the third case, both 'direct' options are (like randomizing) equivalent to a 50–50 gamble. But then why pay for this gamble when you could get it for free?

[67] The argument seems to originate with Savage (1972: 162–4). Joyce (2018: 155–6) applies it to a case like *Hide and Seek*. Kadane and Seidenfeld (1990: 330) apply it to a case like *The Great Lorenzo*. Icard 2019: 119–20 gives a fully general, formal version.

This argument would be unanswerable if it were true that randomizing sets up a gamble whose possible outcomes *are the direct options themselves*. But this is *not* so.

'Options' here are propositions (i.e. sets of possible worlds). Now in *Hide and Seek* the proposition that you go north *directly* is a set of possible worlds at *most* of which the bison went south.[68] But the proposition that you go north *on the augur's advice* is a set of worlds at only *half* of which the bison went south. These two sets of worlds differ in ways that affect their news values.

(Of course, you know that going north directly has the same *effects* as doing so on the advice of the augur. But to assume that options are relevantly alike if known to have the same *effects* is to beg the question against any non-causal decision theory.)

Similarly in *The Great Lorenzo*. The proposition o_1 that you directly pour milk first into teacup no. 1 is a set of worlds. Within this set, those where Lorenzo is fraudulent and chooses that cup get 24% of the credence of the whole set. The proposition r_1 that the robot pours the milk first into teacup no. 1 is another set of worlds. Within that set, the worlds at which Lorenzo is a fraud and chooses teacup no. 1 – those worlds get only 20% of the credence of the whole set. Again these sets differ in ways that matter.

By contrast, we can imagine randomizing options whose outcomes do *not* relevantly differ from the options they ostensibly randomize between. Suppose you must choose one from a set of lottery tickets numbered 1 to 100 with only one winner. You could choose any one directly, or you could use a random number table to choose. The outcomes of *this* randomizing option are indeed the 'direct' options. Suppose you have confidence c_i that ticket i is the winner. The proposition that you choose ticket i directly is a set of worlds in which those where you win get c_i of the credence that you give the whole set. But so is the proposition that you select it after consulting the table. Here randomizing sets up a gamble whose outcomes *are* the options it is randomizing between. Let us call such a process a **pure randomizing process**; the others, where some outcomes are relevantly different from any direct option, are **impure** randomizing processes.

The Bayesian argument only works against pure randomizing processes. The argument was that if there is an optimal direct option, mixing *it* with an inferior option cannot be optimal. Indeed not: but that only shows that no *pure* randomizing process can improve on every direct option. Thus it wouldn't do any good (by your current lights) to randomize your choice of lottery ticket rather than

68 More accurately: a set of worlds *your credence over which* is concentrated on worlds where they go south.

taking one directly; and it would do harm if for some reason you were especially confident about the chances of one particular ticket. But the randomizing processes in *Hide and Seek* and *The Great Lorenzo* are *im*pure: appearances to the contrary, they are not mixtures of the direct options available in those problems but are rather gambles whose possible outcomes are 'acts' that are not directly available at all.[69]

(b) Alternatives to randomization. A second objection concedes that randomization delivers the benefits I mentioned, but denies that it is the *only* way to get them. The following statement of it (by Worrall) concerns clinical trials where an experimenter must assign each subject to the treatment group or to the placebo group.

> If clinicians involved in trials are allowed to decide the arm of the trial to which a particular patient is assigned, then clearly it becomes possible that they will effect, perhaps subconsciously, a selection that distorts the result of the trial and hence gives an inaccurate view of the efficacy of the treatment ... [Eliminating this bias] does seem to me ... a definite epistemological good that properly performed RCTs [i.e. Randomized Controlled Trials] deliver. Notice, however, that there is of course nothing magical about the role of the coin-toss or random number table: selection bias is eliminated in an RCT because the procedure of random allocation means that the experimenter cannot affect the arm that particular patients are assigned to; if the trial was, and remained, double-blind then randomization could play no further role in this respect; moreover even in trials where the experimenters are not blind, any other means of taking the division into experimental and control groups out of their hands would eliminate this potential bias equally effectively.[70]

Indeed there is nothing magical about coins or random number tables, or augury or robots. And what randomization gets us can be got in other ways, not only in clinical trials but also in the cases I discussed. And EDT endorses these other ways, too; but the causal orthodoxy is again committed to rejecting them.

I'll make this point for *TGL*. Adapting that example to Worrall's description: the teacups are the 'patients' and milk-first is the 'treatment'. The effect we are testing for is: being chosen by Lorenzo. Your direct assignment of a 'treatment' to a 'patient' then risks just the bias Worrall describes; and as he says and as I argued, randomization is one way around it.

But there are other ways around it. 'Double-blinding' is one: making you unaware which teacup is getting the milk-first treatment.[71] You might do it like this: there are

---

[69] My strategy here parallels Broome's response to a famous example of Diamond's. See Diamond 1967 and Broome 1991: 111–6.

[70] Worrall 2007: 453–4.

[71] 'Double' blind is misleading: there is no question of the patients' (i.e., the teacups') knowing their treatment.

three switches in a room from which the cups are invisible. Next to each cup is a small milk dispenser which exactly one switch controls, but you cannot tell which switch controls which dispenser. You throw exactly one switch, resulting in the pouring of milk into one cup. Your accomplice notes which cup it is, then pours tea into all three. You then throw the other two switches, pouring milk into the remaining cups. Your accomplice leaves. The Great Lorenzo enters, and so on.

This is not *randomization*, but from EDT's perspective it is just as good. It still breaks any dependence (by the lights of your credence) between which cup gets milk first and which cup Lorenzo identifies – at least any that does not arise from the causal connection you are testing. More precisely: given that Lorenzo is a fraud and cup i gets milk first by this procedure, your confidence that he chooses cup i is 1/3, just as if you had randomized.

Or we might emulate randomization by taking the choice of cup out of your hands and into those of a suitable accomplice. 'Suitable' is doing some work: the accomplice must (you think) be immune to any subconscious influence to which Lorenzo might also (consciously or subconsciously) respond. If she *is* immune then things are epistemically as good as if she were choosing at random, even if she *isn't* random in any plausible sense. Your accomplice might always take the leftmost from any visibly arranged objects of choice; you might even know this, so you can predict which cup gets the milk-first treatment. But if you are certain that Lorenzo has no inkling of it, you will still have confidence 1/3 that he chooses correctly given that he is a fraud, just as if you had randomized.

But the causal approach sees no more value in these *non*-random methods than in randomization. The *effect* of directly choosing teacup no. 1 for the milk-first treatment is the same as the effect of its getting that treatment via the switch mechanism or the accomplice. More generally, because it ignores any non-causal bearing that *your* direct choice of teacup has on Lorenzo's choice, the causal approach always prefers at least one direct choice to randomizing; therefore it also prefers that direct choice to any substitute for randomization that does equivalent work.

It should be obvious that everything I said here regarding *TGL* carries over to *Hide and Seek*. Instead of consulting an augur one might equally use blinding, or an accomplice not subject to subliminal influences. As with *TGL*, EDT would endorse these methods for the same reason it endorses randomization. But the causal approach would reject them for the same reason it rejects randomization.

So my argument is not that EDT is responsive and the causal approach is not responsive to any benefits that only randomization can bring. (There may be no benefits that only randomization can bring.) It is that only EDT is responsive to the benefits of randomization, irrespective of whether randomization or something else is actually bringing them.

5 Harmony

So far, we have considered two distinctive features of EDT: its willingness to use non-causal channels of information (Section 3) and its willingness to block them (Section 4). A third feature is a certain harmony in its treatment of present and future options. Mere displacement from present to future should not affect the value of an option. Nor does it, according to EDT; but as we shall see, the causal orthodoxy violates even a mild version of this principle.

The underlying reason is that the causal orthodoxy sees a fundamental difference between *present options* and everything else. It compares your present options by comparing their effects; but it treats your future options like the choices of somebody else, or like states of nature: things that happen to you, not things you do. So whilst an option is future, its value is its news value; but when it becomes presently available, its value is its causal utility. As we shall see, this discontinuity creates a violent intertemporal disharmony.

EDT by contrast ranks options present or future by their news value. It recognizes no relevant distinction between propositions about what you currently control and those about anything else: all are potential news items. In that sense it treats *everything*, including your choices, as something that just 'happens to you'. A. N. Flew once described Hume's vision of our predicament as a 'paralytic's eye view': there is no 'you' that somehow 'intervenes' on the succession of events: there is only the succession itself, in which some events happen to be 'your' thoughts and feelings, and others 'your' bodily movements, that typically succeed thoughts and feelings of certain types.[72] EDT is Humean too, in that it ignores what this Humean vision rejects. 'Paralytic' is certainly unfair, since neither Hume nor EDT is denying that human actions exist but only saying what they are or what about them matters for choice; but admittedly there is something unsettling about this vision. Still, a better test of a metaphysical picture than how it makes you feel is how it plays out in practice; so let me turn to that.

5.1 The Hamlet Principle

Imagine a choice that you can make now or put off to tomorrow. Putting it off is free, and the parameters of the decision problem you would face tomorrow (the values of outcomes, their probabilities, etc.) are the same as you face today. And you are not going to change your tastes or learn anything relevant between today and tomorrow. I'll call these inert kinds of puttings-off **delays**.

[72] Flew 1954: 49–50; for the Humean vision see Flew *ibid.* n. 13; also Hume 1738 I.iii.14, which rejects a special experience of agency over my current choices, and I.iv.6, which rejects a non-composite self.

Delay cannot be strictly worse than *all* your other present options. The worst you could do by delaying cannot be worse than the worst you could do *now*. Let us call this claim the

Hamlet Principle (HP): For any choice situation, not every other option is better than delay.

The Hamlet Principle looks uncontentious. It isn't saying that you should delay all decisions, nor that it never does harm to delay, nor even that it sometimes does good. It says only that delay is never worse than *all* the other options, not when delay is free.

But CDT violates the Hamlet Principle in the following case.

Optional Delay (OD): Here are two boxes, A and B. We put $1 million in one and nothing in the other. You now have three options.

[a] Take Box A now.
[b] Take Box B now.
[c] Wait and choose in five minutes.

There are two possibilities about where the money is:

[s_A] The $1 million is in Box A.
[s_B] The $1 million is in Box B.

You might think you are choosing completely at random. In fact, subconscious cues in your environment incline you towards one of these boxes (now or in five minutes) with probability $z > 0.5$. We are aware of these cues and used them to place the $1 million. If we detected that the cues would incline you towards Box A, then we put the $1 million in Box B. If we detected that they would incline you towards Box B then we put the $1 million in Box A.

Before you decide, note that delaying is not correlated in any way with the presence, direction or strength of the environmental cues. Nor is it correlated with the whereabouts of the $1 million. You are currently 50–50 as to where the $1 million is. Do you take Box A now, take Box B now, or delay?

The values of the outcomes and the relevant probabilities are as follows: The first two rows of Table 5.1 hold for obvious reasons. The third row holds because if you delay and the $1 million is in Box A, then the environmental cues

Table 5.1 Optional Delay: values of outcomes

	s_A	s_B
a	1	0
b	0	1
c	$1 - z$	$1 - z$

Table 5.2 Optional Delay: values of outcomes

	s_A	s_B
a	$1 - z$	z
b	z	$1 - z$
c	0.5	0.5

incline you to take Box B with probability z; so delaying gives you a probability $1 - z$ of making $1 million. Delaying when the money is in Box A is therefore equivalent to a lottery for $1 million in which your chance of winning is $1 - z$. Similarly when the money is in Box B.[73]

As for Table 5.2, if you start out 50–50 about where the $1 million is, then choosing Box A directly gives you confidence z that you chose the wrong box (i.e. that the $1 million is in Box B).[74] Choosing Box B directly gives you confidence z that the $1 million is in Box A. Delay tells you nothing about what cues are in your environment: so it leaves you 50–50 about where the $1 million is.

It's simple to calculate the news values of your present options, and what EDT recommends on this basis is intuitive:

$$V(a) = V(b) = V(c) = 1 - z.$$

EDT is indifferent between your options and pessimistic about all of them: whatever you do, it reckons that you are unlikely to get $1 million. But it satisfies the Hamlet Principle. It doesn't strictly prefer *both* direct options to delaying the choice between them.

We calculate the causal utilities using Table 5.1 and the facts that (i) $Cr(s_A) = Cr(s_B) = 0.5$ and (ii) which of s_A and s_B is true is causally independent of your present (or any future) decision. But what CDT recommends is not intuitive:

$$U(a) = U(b) = 0.5$$

$$U(c) = 1 - z < 0.5.$$

[73] If a^* says you take Box A after five minutes, b^* that you then take B: $V(a^* \cap s_B) = V(b^* \cap s_A) = 0$ and $V(a^* \cap s_A) = V(b^* \cap s_B) = 1$. And by (2.7) $V(c \cap s_A) = Cr(a^*|c \cap s_A)V(a^* \cap s_A \cap c) + Cr(b^*|c \cap s_A)V(b^* \cap s_A \cap c) = Cr(a^*|c \cap s_A)V(a^* \cap s_A) + Cr(b^*|c \cap s_A)V(b^* \cap s_A) = Cr(a^*|c \cap s_A) = 1 - z$.

[74] $Cr(s_A|a) = \frac{Cr(a|s_A)Cr(s_A)}{Cr(a|s_A)Cr(s_A)+Cr(a|s_B)Cr(s_B)} = \frac{0.5(1-z)}{0.5(1-z)+0.5z} = 1 - z.$

Causal Decision Theory violates the Hamlet Principle. It recommends taking *either* direct option over waiting.[75]

This recommendation is counter-intuitive, but it is intuitive that CDT makes it. It is not some incidental bug that would be absent from a more cautious formulation of the causal approach. It is intrinsic to the causal approach that the contribution of a future option p to its evaluation of a present option o may diverge from its evaluation of p when the latter becomes a present option.

Formalities aside, the reason is as follows. For the causal approach, the merit of choosing a present option o depends on the value of its likely effects: more specifically, on the value of its likely effects in the circumstances in which it is likely to have those effects. If there are environmental cues prompting you to take Box A, then the likely effect of delay is that you take Box A. If there are environmental cues prompting you to take Box B then the likely effect of delay is that you take Box B. Either way, the likely effect of delay is that you miss out on the $1 million. And if your initial credence is (as is natural) 50–50 between s_A and s_B, then the causal merit of *each* 'direct' option exceeds that of delaying. Violation of the Hamlet Principle is intrinsic to the causal approach.[76]

[75] This argument assumes that $\{s_A, s_B\}$ is suitable for $\{a, b, c\}$ (see Section 2.8). It is true that neither $c \cap s_A$ nor $c \cap s_B$ settles *everything* that matters: it also matters which box you ultimately choose. But s_A and s_B do make c evidentially *irrelevant* to everything that $c \cap s_A$ or $c \cap s_B$ fails to settle and that you care about. More precisely, if a^+ says that you ultimately take Box A (i.e. now or later), and b^+ that you ultimately take Box B, the conditions of the problem dictate $Cr(a^+|s_A \cap c) = Cr(a^+|s_A)$ and $Cr(a^+|s_B \cap c) = Cr(a^+|s_B)$, and similarly for b^+. Since settling which of a^+ and b^+ is true, and settling which of s_A and s_B is true, jointly settle everything that matters to you about this problem, it follows that $\{s_A, s_B\}$ is suitable for c by the disjunctive definition at n. 32.

[76] Rothfus (2020) proves that EDT takes a harmonious view of present and future decisions in a wide class of cases that includes *OD* (i.e. those where the subject learns only what her successive choices tell her). But he shows that EDT may exhibit intertemporal *dis*harmony outside this class. Consider a version of *Newcomb's Problem* where both boxes are transparent: those predicted to take both boxes see nothing in the first box and $1,000 in the other, whereas those predicted to take only the first box see $1 million in it and $1,000 in the other. In Rothfus's story, you first decide *whether* to face this transparent version of *Newcomb's Problem*; if yes, you then face it. At the first stage, the second stage option that maximizes news value is taking only the first box, because at the first stage this strongly indicates that you'll make $1 million. But at the second stage (and whatever you then see) EDT recommends taking both boxes. So EDT takes different attitudes to the second-stage decision depending on whether it is present or future.

The difference between this disharmony and the kind for which I upbraided CDT is that the latter responds to *mere* displacement in time, whereas in Rothfus's example what matters to EDT is not just the futurity of the second-stage decision but the new information available in it (i.e. that gained from seeing what money is in the boxes). So the disharmony that he (correctly) highlights does not threaten the Hamlet Principle, which only insists that (information-free) *delay* is never guaranteed to make things worse. Besides, it isn't damning that EDT takes a new attitude when you learn the answer to a question – not even when (as here) the new attitude is the same, *whatever the answer is*. We know from Simpson's Paradox (n. 22) that one can coherently have a preference that future evidence is guaranteed to reverse (see my 2014a: 201).

Still, you might object that it would be possible to *exploit* EDT in a Rothfus-type case, just as it is possible to exploit CDT in *OD2*. For instance, a follower of EDT could be induced to pay, at the

Such violations might be counterproductive (not just counter-intuitive). Consider:

Optional Delay 2 (*OD2*): Like *OD*, but there is a \$100 bonus if you delay.

In *OD2*, EDT prefers delaying to both 'direct' options. You are confident that whichever box you ultimately choose will be wrong, delay or no delay, because you are confident that we have acted on the cues that influence your choice. In these circumstances you might as well take the free \$100.[77]

It is also easy to see that the causal approach still recommends taking either box now over delaying. Since you are 50–50 about where the \$1 million is, the causal approach treats each direct option as a 50% chance of winning \$1 million. But as we saw, it treats delaying as a gamble with a much *less* than 50% chance (a $1 - z$ chance) to win \$1 million. The \$100 bonus won't be worth even a small reduction in your chance of winning \$1 million. So the causal theory recommends turning down \$100 for the sake of acting now.[78]

Now it may be plausible that if you are confident (for whatever reason) that the \$1 million is in Box A, then you ought to go directly for Box A. After all, if you delay, you might change your mind, and it is perhaps reasonable to think that the \$100 bonus for delaying is not worth that risk. But if you are completely uncertain (i.e. 50–50), then it is unclear what harm anyone should see in delay. Maybe delaying would cause you to choose differently from how you would have chosen had you not delayed. But if you are 50–50 about where the \$1 million is, why does that matter? Precisely because of the special status it accords present decisions, the causal approach insists that it does matter, thus giving up \$100 for nothing.

I say 'for nothing' because those who follow this advice can expect to win the \$1 million no more often than those who ignore it; that is, all parties will win \$1 million on the proportion $1 - z$ of occasions when we make a wrong prediction. In consequence, followers of EDT make on average $(1 - z)M + 100$ dollars per trial, whereas followers of the causal approach

first stage, to bind herself to one-boxing at the second stage; in doing so, she is paying to get an outcome that she could have got for free. But in this scenario, the EDT-following agent is *not* gratuitously losing money *in the problem that she is facing*: a problem in which she knows that she will one-box only if she self-binds. In *that* problem, EDT identifies the optimal choice. (Huw Price and I make similar points about another case in our 2012; see also my 2020.)

[77] Suppose for simplicity that the value of the extra \$100 at any level of wealth is x where $0 < x \ll 1$. If a^* says that you take Box A after five minutes, and b^* that you take Box B after five minutes, we have $V(a^* \cap s_B) = V(b^* \cap s_A) = x$ and $V(a^* \cap s_A) = V(b^* \cap s_B) = 1 + x$. So $V(a) = V(b) = 1 - z$ as before, and $V(c) = 1 - z + x$.

[78] In the notation of n. 77, $U(a) = U(b) = 0.5$ as before, and $U(c) = 1 - z + x$: CDT prefers both direct options to delaying, if $z > 0.5 + x$, which is plausible enough given the amounts involved.

make on average just $(1 - z)M$ dollars. Given this clearly superior performance, it is fair to ask the latter: 'If you're so smart, why aren't you rich?'[79]

5.2 The Metaphysical Asymmetry

It would be superficial to criticize its rejection of the Hamlet Principle without addressing the philosophical pressure that drives the causal approach (and tempts everyone) in that direction. This is the fact that decision-making is (isn't it?) obviously about choosing between *futures*, not between pasts.

This arises in the formalism of CDT as follows. When comparing present options, CDT associates a probability distribution with each. The causal utility of an option is then the expectation of the value of the outcome relative to the associated distribution. These distributions agree over the past. They only disagree over the future. And the distribution over the future associated with a particular option reflects your opinion of the future given that option plus your fixed distribution over the past.[80]

If, for example, in *Newcomb's Problem* you are 50–50 over whether there is \$1 million in the opaque box, CDT associates with both one-boxing and two-boxing a probability distribution that assigns 50% to the proposition (about the past)

---

[79] This phrase, and the associated argument against CDT, arose in connection with *Newcomb's Problem* (Lewis 1981b). In that context defenders of the causal approach might plead worse *opportunities*: on this view two-boxing nets a worse average return than one-boxing, not because one-boxing is rationally superior but because there was \$1 million there for most one-boxers to take, but not for most two-boxers (Joyce 1999: 151–4). This response is unavailable in *OD2*: the \$1 million prize and the \$100 bonus are both available to anyone who faces it.

Another example that invites the argument is *The Semi-Frustrater* (Spencer and Wells 2019). This resembles *Hide and Seek* (Section 4.1) except (i) there is no option to randomize; (ii) you can trek in either direction on foot *or* on horseback, but travelling on foot carries a small cost in time; (iii) the bison are more sensitive to the cues that influence those who ride, so it is (you think) more likely that they evade you if you ride than if you walk. Here EDT advises walking and CDT recommends riding. The time gained by followers of CDT does not compensate the high proportion of them that go hungry; the higher success rate for followers of EDT *does* compensate for the extra time. Joyce responds (2018: 152–5) that those that ride are more likely to go hungry whether they ride *or* walk, and so are at a large initial disadvantage compared to those that walk. And this excuses their inferior performance. Whatever the merits of that reply to Spencer and Wells, it too is unavailable in *OD2*. Subjects who directly choose a box have at the outset just as good chances of making \$1 million as those who delay. *Their* inferior performance is more plausibly traced back to their own choices than to any prior state or distribution that they could not have done anything about.

Finally, Wells has recently published a case where followers of EDT seem foreseeably to end up doing worse on average than followers of CDT (Wells 2019). My 2020 responds to this; for other discussions that are more sympathetic to CDT, see Bales 2018 and Gallow 2021.

[80] For the avoidance of doubt, I should stress that the probability distribution that CDT 'associates' with an option o is *not* 'epistemic': it does not reflect your opinion about the past (or about anything else) either before or after you have chosen o. It is rather a constructed distribution that plays a special role in CDT: it is the distribution relative to which the expectation of V gives the causal utility of o. Lewis calls this the distribution that comes from *imaging* on o (1981a: 318).

that the being put $1 million in the opaque box yesterday. So it associates with one-boxing a 50% probability of making nothing and a 50% probability of making $1 million, and with two-boxing a 50% probability of making $1,000 and a 50% probability of making $1 million + $1,000. That is the sense in which CDT treats any choice between options as effectively between the *futures* associated with them.[81]

EDT by contrast compares possibilities that differ over the past. This comes out in the fact that the probability distributions it associates with different options may disagree over the past as well as over the future.

If, for example, in *Newcomb's Problem* you are 50–50 about what is in the opaque box, but 90% confident that the prediction was correct, EDT associates with one-boxing a distribution that gives 90% probability to the proposition that the being put $1 million in the opaque box yesterday. But it associates with two-boxing a probability distribution that gives only 10% probability to this proposition. So it sometimes treats a choice amongst options as a choice between different *pasts* as well as different futures.[82]

The causal approach violates the Hamlet principle because of a general version of the asymmetry in CDT. Informally: given what you think about the actual past, it associates with each option a future that you would expect to be caused by that option together with that past. So choosing a direct option may interfere with the causal impact that you expect the past to have on the future. For instance, in *OD* the causal approach associates with option *a* a 50% probability of breaking the usual influence of environmental cues on your choice of box. But delaying means 'letting nature take its course': in *OD*, the causal approach associates with delay a prospect where the past has the effects that you would expect. This means associating with delay a high probability (z) that pre-existing environmental cues have their usual influence on your future choice. And if those influences are unhelpful, it is not surprising that the causal approach may prefer that you interfere with them by acting now.

[81] More formally: suppose you are choosing from $\omega = \{o_1, o_2 \ldots o_n\}$ and let there be specifications $k \in K$ of everything causally independent of your choice. For any proposition o consider the function f^o taking any proposition p to $f^o(p) = \Sigma_{k \in K} Cr(k)Cr(p|ok)$. f^o is a probability function. The causal utility of any $o \in \omega$ is then the expectation of the news value relative to f^o, that is, given a set Z of outcomes, $U(o) = \Sigma_{z \in Z} f^o(z)V(z)$. So we can regard f^o as the probability function CDT associates with o. Now for any p concerning the past, p is (we suppose) causally independent of your choice, so for each $k \in K$ either $k \subseteq p$ or $k \cap p = \emptyset$. Therefore, for any options o_i and o_j we have $f^{o_i}(p) = f^{o_j}(p) = \Sigma_{k \subseteq p} Cr(k) = Cr(p)$; that is, CDT associates with all of the options probability functions that agree over the probability that they assign to the past.

[82] In terminology parallel to that introduced at n. 81, we can say that the probability distribution that EDT associates with any o chosen from an option set ω is simply Cr_o, because the news value of that option is simply the expected value of the outcomes $z \in Z$ relative to that distribution, that is, $\Sigma_{z \in Z} V(z)Cr(z|o)$ on the assumption that $V(oz) = V(z)$ for any $o \in \omega, z \in Z$.

In short, by holding the past fixed, the causal approach treats your present choice as an intervention on the course of nature rather than part of it. Inevitably then, it sees a *present* option differently from any future copy of it. The upshot is that it attributes to mere delay – mere temporal displacement – a totally implausible ethical significance.

5.3 Deliberating about What Is Past

Implausible it may be, but isn't my preferred alternative absurd? In fact, once it is clear what alternatives the 'evidential' decision-maker is meant to compare, it also becomes clear that people have thought it absurd for millennia.

> We do not decide to do what is already past; no one decides, e.g. to have sacked Troy. For neither do we deliberate about what is past, but only about what will be and admits [of being or not being]; and what is past does not admit of not having happened. Hence Agathon is correct to say 'Of this alone even a god is deprived – to make what is all done to have never happened.'[83]

But isn't this what EDT demands – deliberation about the past, as though we *could* undo 'what is all done'? So however plausible the Hamlet Principle is, and however attractive the intertemporal harmony that makes EDT comply with it, doesn't common sense force us to drop them?

There are two things that Aristotle is saying we (or God) can't, don't, won't or mustn't do. These are (i) to make undone 'what is all done' and (ii) to deliberate about the past. But (i) itself has two readings.

On one reading of (i), Agathon is quite right. It could never be true *both* that something has happened *and* that God or anyone else makes it not have happened. (It could never be true *both* that Troy was sacked *and* that you somehow bring it about that Troy was never sacked.) On this reading, not only is what (i) says about the past true about the past: what (i) says about the past is true about the future. It could never be true both that something *will* happen and that you make it *not be going* to happen. (It could never be true both that I will win big at poker next week and that someone now stops me from winning big at poker next week.)

But it doesn't follow from the fact that no one can do that impossible thing that you can't or shouldn't do (ii) by comparing prospects that differ over the past, just as EDT asks. It no more follows about the past than it follows about the future. We agree that it can never be the case both that something is going to happen and that you somehow make it not happen. But it doesn't follow that you can't or mustn't choose by comparing prospects that differ over the *future*.

[83] Irwin 1985: 150–1 (1139b).

But then equally: even if God cannot *undo* the past, it doesn't follow that He, or we, can't or mustn't choose by comparing prospects that differ over the past.

But (i) is less obviously impossible if Aristotle means that the past *admits* of having not happened, in the sense that if something has happened, then it's *possible* that it didn't happen. The difference between denying (i) on the first reading and denying (i) on the second parallels that between 'Nobody could die on land and at sea', which is obviously true, and 'Nobody who dies on land could have died at sea', which is *not* true (e.g. of a sailor who narrowly escapes a shipwreck only to cast up on an island of killer zombies).

Similarly, on this reading Aristotle means that the past is necessarily the way it is, *not* in the sense that Troy could not have been both sacked and unsacked, but in the sense that if Troy *was* sacked, then it is impossible that it never was sacked. And you can (and it seems he does) say this about the past *without* saying it about the future. That is: one might hold that if Troy *was* sacked, then it *could not* have been unsacked; but also that I will in fact win big at poker next week even though it *could* now have been the case that I won't.[84]

The pointlessness or impossibility of (ii) follows from the impossibility of (i) on this second reading. If the past does not admit of being otherwise and the future does, there may be no point deliberating about the past even if there is a point to deliberating about the future. However the past was, it *now* cannot have been otherwise; but however the future will be, it now *might* be otherwise. If (in *Newcomb's Problem*) you are comparing a one-boxing world where you get $1 million with a two-boxing world where you only get $1K, then you are comparing worlds at least one of which is now impossible. Knowing this should make such deliberation pointless.

But there are as many meanings of 'can' as there are sets of facts. What gives this one a special say in deliberation?

> To say that something can happen means that its happening is compossible with certain facts. Which facts? That is determined, but sometimes not determined well enough, by context. An ape can't speak a human language – say, Finnish – but I can. Facts about the anatomy and operation of the ape's larynx and nervous system are not compossible with his speaking Finnish. The corresponding facts about my larynx and nervous system are compossible with my speaking Finnish. But don't take me along to Helsinki as your interpreter: I can't speak Finnish. My speaking Finnish is compossible with the facts considered so far, but not with further facts about my lack of training. What I can do, relative to one set of facts, I cannot do, relative to another, more inclusive set.[85]

[84] Pasnau 2020: 236ff. discusses a parallel ambiguity in a famous passage in *De Interpretatione*.
[85] Lewis 1976: 77.

In general, for any set of facts – any proposition – there is a corresponding notion of possibility: what is impossible relative to those facts is what they rule out, where 'rules out' may be interpreted as some nomological kind of incompatibility (as Lewis seems to intend), or as 'logical' or 'metaphysical' incompatibility.

Relative to all facts – known and unknown – about the *past*, either there cannot have been $1 million in the opaque box in *Newcomb's Problem* or there cannot have been nothing there. Similarly in *Beer or Soda* (Section 3.2), relative to all past facts it is not both: possible that at the time of deliberation you have a weakness for alcohol *and* possible at that time that you do not. In *The Great Lorenzo* (Section 4.2) it is not possible that subconscious cues encourage you to choose teacup no. 1 for getting the milk-first treatment whilst also being possible that they encourage you to choose no. 2. Relative to all facts about the past, alternatives that differ over the past are never compossible.

But relative to other sets of facts they are. In particular: relative to the facts – past, present and future – that the deliberating subject knows, they are all possible. You don't know whether you have a weakness for alcohol: relative to what you do know, it's possible you do and possible you don't. You don't know whether some subliminal variation in texture, colour or brightness exists that encourages you to assign the treatment to teacup no. 1. Relative to what you know this is possible; also possible that this variation inclines you towards teacup no. 2 and possible that it inclines you towards teacup no. 3. Relative to what you know (or to those things of which you are certain, or to those of which you are certain enough), alternatives that differ over the past often *are* compossible.

So of all the kinds of possibility that there are, why think it is possibility relative to facts about the – known or unknown – past that settles what you can sensibly deliberate about? Why not, for example, possibility relative to what you know?[86]

'Well, the past is out of your causal reach. *That* is why there is no point in deliberating about it.'

The past may not be out of your causal reach.[87] But it's plausible enough, in *Beer or Soda* and *Hide and Seek*, that what matters about the past (e.g. where the bison went) *is* out of your causal reach. Still, this is no argument. The attack on deliberation about the past was directed at EDT, which denies from the outset

[86] List 2014 argues similarly for kinds or 'levels' of possibility, distinguishing the macro-level at which one could have acted otherwise from the micro-level at which determinism implies that nothing could have been otherwise. On this view, the possibilities that you could realize by choosing otherwise will differ in their micro histories as well as their macro futures. Cf. Loewer 2007 and my 2014a §5.4.

[87] Dummett 1964 is a classic defence of backward causation. For a recent defence that explicitly connects with present concerns see Price and Liu 2018.

that its causal influence on the outcome is what matters about an option. On the contrary, EDT focuses on its evidential bearing, which can reach beyond its sphere of merely causal influence. Just asserting that deliberation should ignore what you cannot *affect* is not going to persuade anyone.

But neither Aristotle, nor to my knowledge anyone else, offers any other reason for giving possibility-relative-to-the-past this special status vis-à-vis practical deliberation. (Or is there meant to be evidence for this, for instance, empirical evidence, that partisans of EDT have somehow overlooked?) Aristotle is wrong if he means that we shouldn't – and wrong in any case to suggest that no-one ever does – deliberate about the past. In *Newcomb's Problem*, and in the more realistic problems that this Element has discussed, people do just that; and for all we have seen there is nothing wrong with it.[88]

6 Conclusion

This portrait of EDT described its distinctive applications and contrasted it with the causal orthodoxy. As you can imagine, there are many other points worth covering regarding both its applications and its contrast with other theories. This conclusion sketches two.

There is a kind of ethical subjectivism built into the heart of EDT. It can be shown to entail that there is no such thing as *objective conduciveness* to your ends. There is no feature of a proposition that constitutes its real or objective conduciveness to your aims.

The reason (in short) is that for any objective function that assigns a number $F(o)$ to an option o, we may construct a Newcomb-like situation offering a choice between o_1 and o_2, such that for some partition $\{s_1, s_2\}$:

- If s_1 is true, then $F(o_2) > F(o_1)$ and both are high.
- If s_2 is true, then $F(o_2) > F(o_1)$ and both are low.
- o_1 is strong evidence that s_1 is true.
- o_2 is strong evidence that s_2 is true.

[88] Another, more EDT-friendly, argument against deliberation about the past goes like this: when you are deliberating, you are aware of desires or other mental states by which the past influences your decision, and these states 'screen off' your choice from any past state i.e. make the former evidentially irrelevant to the latter. So your options have no evidential bearing on the past, and there is no *need* to deliberate between alternatives that differ over the past (Eells 1982). Some philosophers have used this point in an argument that realistic versions of *Newcomb's Problem* never arise (see Bermúdez 2018). Other defenders of EDT have argued that even if you are *un*aware of the evidence that you are determined to act in this or that way, the objective evidence that screens off your present choice from its non-effects is still there and still available to you (Fernandes 2017: 703–4). But in *Beer or Soda* it is plausible that your present desires or other mental states – at least those available without MRI technology – do *not* exhaust the evidential relevance of your choice to your tendency to do similar things in similar cases. The best present predictor of what you do in the future is not what you now think or want, but what you now *do*.

Here *you know for sure* that o_2 has a higher F-score than o_1, because o_2 has a higher F-score than o_1 given either s_1 or s_2, one of which is certainly true. So, if F measures conduciveness to your ends, then you know for sure that o_2 is more conducive to your ends than is o_1. But EDT advises you to choose o_1, because doing so is evidence that your chosen option is *highly* conducive to your ends, whereas choosing o_2 is evidence of the opposite.

So if F measures objective conduciveness to your ends, EDT recommends an option that you *know* is less conducive to your ends than some alternative. It follows that if EDT is true, then F cannot measure objective conduciveness to your ends: no function F from options to numbers could *count* as 'measuring objective conduciveness to ends' unless it satisfies this condition: a rational agent chooses an F-maximizing option if he knows how. But if EDT is true, then in this Newcomb-like situation a rational agent is not choosing the known F-maximizing option o_2. It follows that if EDT is true, then there is no such thing as objective conduciveness to your ends.[89]

This argument separates EDT from a wide range of decision theories, all of which are consistent with objective conduciveness. CDT itself, for instance, determines that relation quite precisely: an option's objective conduciveness to your ends is the strength with which it would in fact cause them to be realized; and many other decision theories, like maximin, certainly allow for such a thing as objective conduciveness to your ends. In essence, the difference is that all these theories rank options by a function that can be interpreted as your estimate of *some* objective quantity. What the foregoing argument shows – or rather what it would show given the necessary details – is that news value cannot be your estimate of *anything*. Quite how destructive this is, either of common-sense morality or of Platonic notions of ethical objectivity, or for that matter of EDT itself, remains to be seen.

A recent challenge to EDT as a normative theory of choice arises not from the causal orthodoxy but from a recently popular alternative paradigm known as *Functional Decision Theory* (FDT). This theory, which has gained currency especially in AI, is a successor to Yudkowsky's 'Timeless Decision Theory'.[90] Here is the idea:

> FDT suggests that you should think of yourself as instantiating some decision algorithm. Given certain inputs, that algorithm will output some action. Because algorithms are multiply realizable, however, there can be other

[89] Ahmed and Spencer 2020. Our conclusion (though not our argument) is related to Lewis's (1988, 1996) rejection of 'Desire as Belief'; in fact, we believe that this defence of Lewis's Humean thesis gets around some objections that were effective against Lewis's original argument (e.g. Price 1989).

[90] Yudkowsky 2010.

> instantiations of this same (or a very similar) algorithm, and, more generally, the state of the world can non-causally depend on what your decision algorithm does. Your decision procedure is not local to your own mind, and rationality requires taking this fact into account.[91]

How best to spell this out is not yet clear, but it is possible to get an idea of some intended applications. FDT evaluates an option *o* by asking what would have happened, not in the counterfactual situation in which the agent *performs o*, but in the counterfactual situation in which the agent's *algorithm would have outputted o* given the agent's current beliefs and desires as input.

There are cases where FDT makes recommendations that differ from both EDT and CDT. The following is particularly vivid.

> *Parfit's Hitch-hiker.* An agent is dying in the desert. A driver comes along who offers to give the agent a ride into the city, but only if the agent will agree to visit an ATM once they arrive and give the driver $1,000. The driver will have no way to enforce this after they arrive, but she does have an extraordinary ability to detect lies with 99% accuracy … In the case where the agent gets to the city, should she proceed to visit the ATM and pay the driver?[92]

Both EDT and CDT advise the agent not to pay. At that point the agent has already been rescued and has nothing to gain from paying the driver *now.* FDT disagrees: instead of asking, *what would happen if I were to pay the driver / not pay the driver now*, it asks: *what would have happened if I had been running an algorithm that would have recommended paying the driver / that would have recommended not paying the driver*? To answer these questions, we go back in time to when your algorithm first makes a difference (i.e. back in the desert). At that point, your running an algorithm that would later have recommended paying the driver would have meant that the driver could see this and so would have given you a lift. But running an algorithm that would later have recommended *not* paying the driver would have meant that the driver could see *this* instead and would therefore have left you in the desert. So the algorithm it would have been best for you to have all along now recommends paying the driver. FDT therefore recommends paying the driver.

It is hard not to find this surprising: after all, you are giving up $1,000 when doing so neither causes nor is evidence of any compensation. But defenders of FDT can point out that followers of FDT typically lose only $1,000 in situations like this, whereas followers of EDT or CDT typically lose much more. In reply to *this*, distinguish two claims:

[91] Levinstein and Soares 2020: 9–10. [92] Soares and Yudkowsky 2018: 8.

You would have been better off running an algorithm that outputs the recommendation to pay than running one that outputs the recommendation not to pay. (6.1)

You are better off paying now than not paying now. (6.2)

Parfit's Hitch-hiker establishes (6.1) but not (6.2); but only (6.2) addresses the question that normative decision theory was supposed to be answering: however things *might* have been, given how they *actually* are, what are you now supposed to do? From this perspective FDT gives a good answer to a different question, one you should perhaps have asked at the outset: what sort of algorithm should I be running on this trek? But once you are safely out of the desert, the time for asking *that* is behind you.

I am therefore inclined not to see FDT as a genuine rival to EDT and CDT but as complementary to them. This may be premature: FDT is a new approach for which many possible lines of development remain open, and a sophisticated and mathematically rigorous version of the theory may force us to rethink some fundamental features of the normative approach.[93]

But as things stand, Evidential Decision Theory represents the most plausible and fully worked-out alternative to the causal orthodoxy. The decision-making self is not a philosophical singularity, somehow lifted free of the natural influences that connect everything to everything else; nor is there any point in pretending it is. On the contrary, it is always true and it often matters that you and your choices are wholly embedded in the causal web of nature; that EDT takes full account of this fact is a distinctive feature of it that speaks decisively in its favour.

[93] For instance, there is currently lack of clarity surrounding the counterfactuals at the heart of FDT: counterfactuals about what would have happened if a given algorithm i.e. a given mathematical object had yielded a different output for a given input. Proponents of FDT admit that these considerations involve the dubious notion of a logically impossible world (see e.g. Soares and Yudkowsky 2018: 8). The trouble is that if we give up on logic, we are hard-pressed to find any principled way to determine what would or would not have been the case at a world. But it is natural to sympathize with Soares and Yudkowsky when they write that this is a 'technical' rather than a philosophical difficulty (pp. 7–8). Similar difficulties arise for Bayesian attempts to represent scientific reasoning. For instance, you might think that the problem of old evidence arises because standard Bayesian norms leave no room for reasoning under conditions of ignorance about logic (e.g. Garber 1983). If so, you might reasonably expect a plausible account of such reasoning to impose a kind of discipline on reasoning that could also be applied to FDT.

Appendix A
Non-binary News Value

This Appendix covers two possible complications of the simplified approach to news value in Section 2.3. It describes two ways to drop the assumption that the agent has only one ultimate value: first (A2) by supposing that she has many ultimate values, and then (A3) by supposing that she has none. Before that (A1) I introduce the basic conative notion of preference that is essential to both developments.

A1 News Preference

You would rather learn some things than others. Planning a picnic for tomorrow, you would rather learn that it will be sunny all day than that there will be lightning. If in a multiple-choice exam you answered 'A' to the first question and your enemy answered 'C', you'd rather learn that the correct answer was A than that it was B, and you'd rather learn that it was B than that it was C.

To learn something – to receive news – is to adjust your credences in response to perceptual or other inputs. Given moderate idealization, such adjustments are often describable as Bayesian updating on a *proposition*. That is, when a learning process shifts your credence from Cr to Cr^*, there is some proposition p such that $Cr^*(q) = Cr(q|p)$ for any proposition q: your credences shift as they would if you had learnt for certain exactly that p is true.

One might doubt whether all learning processes are so describable.[1] Whatever you think about that, it is plausible that some are describable in another way: not as the learning of a *proposition* but as the learning of a *distribution*. What does this mean?

> *Blofeld or Scaramanga?* When James Bond came to, he found himself at an elaborately laid table in an opulent mansion. A waiter pattered in to fill his glass from an unmarked bottle. I must have been kidnapped, Bond thought. Where am I? I know it is either Blofeld's underground lair or Scaramanga's island fortress. But which? Right now, he was 50–50.
>
> The wine was a clue. He knew it was either French or Australian. He knew that Blofeld was partial to Australian wine and Scaramanga to French. So,

[1] Jeffrey (1983: 165) describes seeing a cloth by candlelight: one learns something from this experience about its colour, but there may be no proposition expressible in English that this learning experience can be modelled as conditionalization upon. Christensen (1992: 545) suggests that a perceptual demonstrative – 'I'm appeared to *that* way' – might work in an appropriate context; but as he points out, the propositions that such sentences express would be hard to fit into the Bayesian framework.

learning that it was Australian would make him 80% confident that he was in Blofeld's underground lair, and learning that it was French would make him 80% confident that he was in Scaramanga's island fortress. Right now, he was 50–50 about the wine too.

But when he *drank* it, Bond learnt something. He became more confident that it was French – not certain, but by the time he finished, his confidence that the wine was French had shifted from 50% to 90%.

For any finite partition $G = \{g_1 \ldots g_n\}$, let a **distribution** over G be a function $\Pr: G \to [0, 1]$ such that $\sum_{i=1}^{n} \Pr(g_i) = 1$. To say that **you learn the distribution Pr** is to say that you have an experience that for each g_i directly adjusts your confidence in it to $\Pr(g_i)$ without changing your relative confidence in any of the propositions that entail g_i.

By drinking the wine Bond learnt the distribution $\Pr(f) = 0.9$, $\Pr(\bar{f}) = 0.1$ over the partition $\{f, \bar{f}\}$, f being the proposition that the wine is French and $\bar{f}$ the proposition that it is Australian. Now, the probability that he was in Blofeld's underground lair was 20% given that the wine was French and 80% given that it was Australian. The probability that he was in Scaramanga's island fortress was 80% given that it was French and 20% given that it was Australian. These conditional probabilities did not change. Drinking the wine made him 90% confident that the wine was French and therefore 90% (80%) + 10% (20%) = 72% confident that he was in Scaramanga's island fortress. Bond had updated on a distribution.

More generally, if Pr is a distribution over a partition $G = \{g_1 \ldots g_n\}$, then to update on Pr is for one's credences to undergo **Jeffrey conditionalization**: a shift from Cr to $Cr_{\Pr}$ where, for any proposition q over which Cr is defined,

$$Cr_{\Pr}(q) = \sum_{i=1}^{n} \Pr(g_i) Cr(q|g_i). \tag{A1}$$

And a learning episode is describable as an update Pr if it involves the shift in credences that *would* have occurred if one *had* learnt exactly that the probability of each of the g_i was $\Pr(g_i)$, though it might in fact have occurred because you learnt something else.[2] It is plausible that some shifts in credence are describable as updates on distributions.

But it's *also* plausible that your *preferences over learning episodes* extend to such shifts. Because of your plans for a picnic, you'd prefer learning for sure

[2] Joyce calls these *hard* Jeffrey shifts: any hard Jeffrey shift classes together learning episodes in which the associated distribution over some partition is imposed upon the subject. There are also soft Jeffrey shifts, which class together learning episodes in which the subject's credences over the elements of a partition change by some *ratio*: for instance, all episodes in which your credence in p_1 falls by a half, your credence in p_2 doubles etc. (Joyce 2011; Huttegger 2017 ch. 8 is a useful discussion). I do not know how much of the structures in A.2 and A.3 would be recoverable from complete preferences over soft Jeffrey shifts.

that the sun will shine all day tomorrow over learning for sure that there will be lightning. That is, you'd prefer a learning episode that shifted my credence in sunshine to 1 over one that shifted my credence in lightning to 1. But by the same token, you'd also prefer a learning episode that shifted your credence in sunshine to 0.9 and your credence in lightning to 0.05 over one that shifted your credence in sunshine to 0.4 and your credence in lightning to 0.5.

Let us assume that you have complete and transitive **preferences** over all Jeffrey conditionalizations over finite partitions. That is, for any finite partition $G = \{g_1 \ldots g_n\}$, let a *distribution* over G be a function $\Pr: G \rightarrow [0, 1]$ such that $\sum_{i=1}^{n} \Pr(g_i) = 1$. Let Δ be the set of all such distributions over all such partitions. Then there is a transitive, irreflexive and asymmetric relation $\succ$ on Δ^2 which holds between two distributions d_1 and d_2 (possibly over different partitions) if you prefer learning d_1 to d_2. That is, if d_1, d_2 and d_3 are any three such distributions, then:

- If $d_1 \succ d_2$ and $d_2 \succ d_3$ then $d_1 \succ d_3$ ($\succ$ is transitive)
- Not $d_1 \succ d_1$ ($\succ$ is irreflexive)
- If $d_1 \succ d_2$ then not $d_2 \succ d_1$ ($\succ$ is asymmetric).

If neither $d_1 \succ d_2$ nor $d_2 \succ d_1$, then I'll say you are **indifferent** between d_1 and d_2, which I write $d_1 \sim d_2$.

In the following exposition I'll take these preferences as primitive. Note that the assumption of preference over distributions implies that preference over propositions, because learning a proposition is a special case of learning a distribution: to learn a proposition p is to learn the distribution Pr over $\{p, \overline{p}\}$ such that $\Pr(p) = 1$.

A2 Many Ultimate Values

We can now weaken the assumption that the agent is as single-minded as assumed at Section 2.3. Suppose there is a finite partition $\underline{G}$ of Ω into sets $g_1 \ldots g_n$ which capture everything you care about, in the sense that once you have learnt which of the g_i is true, you have learnt everything that you care about. Formally, I'll call $\underline{G}$ a **value system**, meaning that for any p and q and any $g_i \in \underline{G}$ such that $Cr(p \cap g_i), Cr(q \cap g_i) > 0$, you are indifferent between learning $p \cap g_i$ for sure and learning $q \cap g_i$ for sure.[3]

For instance, suppose you care about *two* things: (e) that England wins the football World Cup in 2066, and (s) that Somerset wins the cricket County Championship at least once. Your value system $\underline{G}$ has four elements:

[3] The results sketched here can also be extended to denumerable and uncountable analogues of value systems: see Kreps 1988: 59ff.

- g_1 : England wins in 2066 and Somerset wins at least once: $e \cap s$
- g_2: England wins in 2066 but Somerset never wins: $e \cap \bar{s}$
- g_3: England doesn't win in 2066 but Somerset wins at least once: $\bar{e} \cap s$
- g_4: England doesn't win in 2066 and Somerset never wins: $\bar{e} \cap \bar{s}$

You know that exactly one of g_1, g_2, g_3 and g_4 is true. What matters to you about any proposition, or about any distribution, is its bearing on which of them is true, that is, how learning it will redistribute your credences amongst g_1, g_2, g_3 and g_4.

I'll make two assumptions about your preferences for learning distributions, in particular distributions over $\underline{G}$. I'll use the following notation. Just as we can add and multiply numbers, so too we can add and multiply functions, in particular distributions with the same domain and a numerical range. If f and g are such functions, then $f + g$ is that function h such that $h(x) = f(x) + g(x)$ for any x lying in the domains of f and g. And if a is a real number, then af is that function g such that $g(x) = af(x)$ for any x lying in the domain of f.

Let $\Delta_{\underline{G}}$ be the set of all distributions or 'lotteries' over your value system $\underline{G}$. You have preferences over these distributions because you have preferences over *all* distributions. Returning to the last example, you might, for example, prefer learning exactly that the probability of g_1 is 0.8 and the probability of g_2 is 0.2 over learning that the probability of g_1 is 0.75 and the probability of g_4 is 0.25. In any case, whatever finite partition $\underline{G}$ is, the assumptions are that for any $d_1, d_2, d_3 \in \Delta_{\underline{G}}$ we have:

> **Independence**: If $x \in (0, 1)$ and $d_1 \succ d_2$ then $xd_1 + (1 - x)d_3 \succ xd_2 + (1 - x)d_3$ (A2)

> **Continuity**: If $d_1 \succ d_2 \succ d_3$ then $xd_1 + (1 - x)d_3 \succ d_2 \succ yd_2 + (1 - y)d_3$ for some $x, y \in (0, 1)$. (A3)

In (A2) and (A3), $xd_1 + (1 - x)d_3$ is the distribution over $\underline{G}$ that you get from a weighted sum of the distributions d_1 and d_3 where the weights are x and $(1 - x)$. For instance: suppose d_1 gives 50% probability to g_1 and 50% probability to g_2, and d_3 gives 70% probability to g_1 and 30% probability to g_3. Then if $x = 0.5$, the distribution $xd_1 + (1 - x)d_3$ gives 60% probability to g_1, 25% to g_2 and 15% to g_3.

These assumptions are intuitive. The independence principle (A2) says roughly that if you prefer learning that you have won a car to learning that you have won a holiday, then you prefer any chance of winning the car to the same chance of winning the holiday. More generally, it says that whether you prefer learning one distribution to another depends only on the propositions over which they differ. This has an abstract plausibility, although there are well-known cases where it fails as a descriptive principle.[4]

The continuity principle (A3) says roughly that if you prefer winning a car to winning a holiday, and you prefer winning a holiday to electrocution, then there is a lottery between the car and the electrocution (with a chance of the latter) that you prefer to the holiday, and one that you'd disprefer to it. The first half of this may seem implausible: wouldn't you turn down *any* gamble with your life? But what if I said: you can have either a holiday or the car of your dreams, but the default is the holiday: if you want the car you must confirm by e-mail. So, getting the car involves operating an electric device (to send an e-mail), which does carry a tiny but non-zero risk of electrocution. But you would still do it.[5]

From (A2) and (A3) follows:

There is a function $u: \underline{G} \rightarrow \mathbb{R}$ such that for any $d_1, d_2 \in \Delta_{\underline{G}}$:

$$d_1 \succ d_2 \text{ iff } \sum_{g \in \underline{G}} g \in \underline{G}\, x_1(g) > \sum_{g \in \underline{G}} u(g) d_2(g). \qquad \text{(A4)}$$

Also, if u' is another function of which (A4) also holds, then there are real numbers x and y such that $x > 0$ and $u' = xu + y$. In this case, u' is called a *positive affine* transformation of u.[6]

We can therefore represent the values of each element in your value system by a number, called its **utility**, in such a way that the value of any distribution over that value system can be represented as the expectation of utility according to that distribution, where 'representation' means that for any two distributions d_1 and d_2, you prefer d_1 to d_2 if and only if it has greater expected utility.

Returning to our example: suppose your preferences for distributions over $\underline{G} = \{g_1, g_2, g_3, g_4\}$ satisfy (A2) and (A3). So there is a u satisfying (A4). Suppose some u satisfying (A4) is such that:

- $u(g_1) = 100$
- $u(g_2) = 60$
- $u(g_3) = 50$
- $u(g_4) = 0$.

Then we can calculate your value for, and hence preferences over, any distribution or lottery over $\underline{G}$. For instance, how pleased would you be to learn the distribution d_1, which gives a probability of 0.8 to g_1 and of 0.2 to g_2? How

---

[4] Allais 1953. For discussion of the normative status of the independence principle, see Buchak 2013 ch. 5; I discuss Buchak's work on this in my 2016.

[5] Kreps 1988: 45 offers a similar example.

[6] Von Neumann and Morgenstern 1953 ch. 3 and appendix.

pleased would you be to learn d_2, which gives a probability of 0.75 to g_1 and 0.25 to g_4? By (A4):

$$\sum_{i=1}^{4} u(g_i)d_1(g_i) = 0.8(100) + 0.2(60) = 92$$

$$\sum_{i=1}^{4} u(g_i)d_2(g_i) = 0.75(100) + 0.25(0) = 75.$$

According to this assignment of value u to the g_i (and according to any positive transformation of it) you will, if you are rational, prefer learning d_1 to learning d_2.

We then extend this valuation to arbitrary propositions. Suppose your value system is $\underline{G}$. Then when you learn a proposition (or any distribution), all you care about is how it distributes your credences amongst the elements of $\underline{G}$. Now when you learn a proposition p, it is a consequence of simple Bayesian updating (see (2.4) at Section 2.2) that your credence in any $g_i \in \underline{G}$ shifts from your original $Cr(g_i)$ to $Cr(g_i|p)$. So you should be just as pleased to learn the news that p as you are to learn the distribution d_p over your $\underline{G}$ that assigns probability $Cr(g_i|p)$ to g_i. This means that the value to you of learning any proposition is just $\sum_{i=1}^{i=n} u(g_i)d_p(g_i)$. We can define this as the *news value* $V^{\underline{G}}$ of the proposition

$$V^{\underline{G}}(p) = \sum_{g \in \underline{G}} u(g)Cr(g|p). \tag{A5}$$

Putting V for $V^{\underline{G}}$, for any propositions p and q such that $Cr(p) > 0$, $Cr(q) > 0$, we have

$$p \succ q \text{ if and only if } V(p) > V(q). \tag{A6}$$

Any V for which (A6) holds is said to **represent** your news preferences (for propositions).

It is easy to see that binary news value is the special case where $\underline{G} = \{g, \overline{g}\}$ for some g such that $g \succ \overline{g}$. It is also easy to derive the fundamental equation for news value: if p is any proposition such that $Cr(p) > 0$ and if $\{q_1, q_2 \ldots q_n\}$ is any partition such that $Cr(p \cap q_j) > 0$ for each q_j, then

$$V^{\underline{G}}(p) = \sum_{j=1}^{n} Cr(q_j|p)V^{\underline{G}}(p \cap q_j). \tag{A7}$$

The proof is as in the binary case.[7]

So, if she has a value system $\underline{G}$, and news preferences for distributions over its cells satisfying (A2) and (A3), the agent's news preferences over arbitrary

[7] Define $\Gamma(p) = V^{\underline{G}}(p)Cr(p)$. Clearly Γ is additive, so if $\{q_1, q_2 \ldots q_n\}$ is a partition s.t. $Cr(p \cap q_j) > 0$ for each q_j, then $\Gamma(p) = \sum_{j=1}^{n} \Gamma(p \cap q_j)$ i.e. $Cr(p)V^{\underline{G}}(p) = \sum_{j=1}^{n} Cr(p \cap q_j)V^{\underline{G}}(p \cap q_j)$; since $Cr(p) > 0$ it follows by (2.4) that $V^{\underline{G}}(p) = \sum_{j=1}^{n} Cr(q_j|p)V^{\underline{G}}(p \cap q_j)$.

propositions can be represented by a news value function $V^{\underline{G}}$ assigning a number to each proposition.

A3 No Ultimate Values

So far, we have assumed that there is some finite set of prizes against which all others could be calibrated. Everything else matters because, and only because, of the subjective 'lottery' that it induces over these prizes. But it might happen, on the contrary, that for any level of detail there are specifications of reality at that level that leave open questions to which you are not indifferent. To cover this, we must develop a usable notion of news value that doesn't assume a value system, and which doesn't involve numerical utilities at all.

Instead, assume only this: that *how pleased you are to learn something depends only on what learning it teaches you.* That is, if each of p and q is a proposition or a distribution on an arbitrary partition, then you should be indifferent between learning either if doing so has the same effect on your beliefs:

$$Cr_p = Cr_q \rightarrow p{\sim}q \tag{A8}$$

This equivalence is plausible because $\succ$ is tracking preferences over learning episodes. But if $Cr_p = Cr_q$, then you must be disposed to learn the same when you learn p as when you learn q. If you preferred p to q or q to p in these circumstances, then $\succ$ would be sensitive to something other than the news that these propositions or distributions bring you.

Now if p is a proposition and $G = \{g_1, \ldots g_n\}$ is any partition, define the **projection of *p* onto *G*** as a distribution over the partition $\{\overline{p}\} \cup \{p \cap g | g \in G\}$, call it π_G^p, such that:

$$\pi_G^p(\overline{p}) = 0$$

$$g \in G \rightarrow \pi_G^p(p \cap g) = Cr(g|p).$$

The following principle then holds for any proposition p and any partition G:

Distributional Equivalence: $p{\sim}\pi_G^p$ (A9)

You are just as pleased to learn that p is true as you are to learn exactly that the probability of $p \cap g$ is $Cr(g|p)$ for each cell g of an arbitrary partition G.[8]

Distributional Equivalence creates a standard on news preferences by which to determine which of two propositions p, q is better news. You proceed as

[8] Proof: let q be an arbitrary proposition. Then by (A1) $Cr_{\pi_G^p}(q) = \sum_{g \in G} Cr(q|p \cap g)$ $\pi_G^p(p \cap g) + Cr(q|\overline{p})\pi_G^p(\overline{p}) = \sum_{g \in G} Cr(q|p \cap g)Cr(g|p) = \sum_{g \in G} Cr(q \cap g|p)$ (by (2.4)). But since G is a partition, this is $Cr(q|p) = Cr_p(q)$. So $Cr_p = Cr_{\pi_G^p}$, so by (A8) $p{\sim}\pi_G^p$.

follows: find suitable partitions G and H such that you *are* sure which of π_G^p and π_H^q you'd prefer to learn. Then by (A9) p is better news than q if π_G^p is better news than π_H^q and vice versa; and if π_G^p and π_H^q are equally good news, then so are p and q.

For instance, suppose you must choose between becoming an astronaut (p) and becoming a bank robber (q). It might not be obvious which of p and q you'd prefer to learn. But suitable partitions could clarify things. For example, let g be the proposition that you get to visit Mars and let h be the proposition that you end up in jail; let the partitions in question be $G = \{g, \overline{g}\}$ and $H = \{h, \overline{h}\}$. Now suppose you are 60% confident that you will get to visit Mars given that you become an astronaut and 80% confident that you will spend time in prison given that you become a bank robber. Then you prefer learning p to learning q if and only if you prefer π_G^p to π_H^q: that is, if and only if you prefer (i) a 60% prospect of visiting Mars and a 40% prospect of being an astronaut who never gets there to (ii) bank robbery with an 80% prospect of time in jail and a 20% prospect of freedom (and riches). Although these preferences are equivalent, comparing these projections of p and q onto G and H respectively might make your preference between p and q themselves more obvious.

Or to anticipate, in *Newcomb's Problem* (Section 2.7) you must choose between (o_1) taking only the opaque box and (o_2) taking both boxes. If s_1 says that the being predicted that you would take only the opaque box, and s_2 says that the being predicted that you would take both boxes, then we can project both options onto the partition $S = \{s_1, s_2\}$. Then $\pi_S^{o_1}$ gives you a 99% prospect of \$1 million and a 1% prospect of nothing (see Table 2.6) whereas $\pi_S^{o_2}$ gives you a 99% prospect of \$1,000 and a 1% prospect of \$1,000 + \$1 million. Obviously you would prefer to learn the former; hence by distributional equivalence you would rather learn o_1 than o_2. There is nothing especially surprising about this result: the point is that we can obtain it without having to assign numerical utilities to any outcome.

In fact, the distributional equivalence suffices for many applications of news value, including all those studied here.[9] We are interested in comparing propositions to determine what is better news than what. We can do that by comparing appropriate projections of those propositions. We thus settle news

[9] None of the applications in this Element exploit the fact that we may use *different* partitions to compare propositions, as in the just-discussed 'astronaut' case. In all the applications we effectively compare the projections of p and q onto a single partition G. The reason for this is to allow for the simpler calculation not of news value but of a contrasting notion that I call causal utility. Calculating the causal utility of options requires specification of the outcome of each choice in each cell of a single, common partition: see Section 2.8.

preferences between propositions without appealing either to a numerical scale of news values or to *any* 'ultimate' values.

Having said that, representing the news value of a proposition by a cardinal number makes it possible to treat news value, and EDT, in simpler and more familiar terms. I will therefore use numerical news values in what follows; indeed I'll take for granted the rather demanding assumptions of Section 2.3. But in principle we could do without any of that.

Appendix B
Randomization and Accuracy

The example in Section 4.2 used the simple distribution of values at Table 4.3. These values are realistic enough. But they might seem naïve to a Bayesian whose response to any experiment is not 'yes' or 'no' but an updated credence. A more sophisticated valuation would record the *expected accuracy* of the resulting probability distribution over the variable of interest, the accuracy being measured by some sort of **scoring rule**.

For present purposes, we can treat a scoring rule as a function f^p that measures the accuracy of your credence in a particular proposition p of interest. More specifically, we can identify the scoring rule relative to p as a pair of functions s_1 and s_0 such that:

$$f^p(Cr) = \begin{cases} s_1(Cr(p)) \text{ if } p \text{ is true} \\ s_0(Cr(p)) \text{ if } p \text{ is false} \end{cases}.$$

For instance, let p be the proposition that it rains tomorrow and let the functions s_1 and s_0 be $s_1(x) = (1-x)^2$ and $s_0(x) = x^2$. Suppose that my credence that it rains tomorrow is $Cr(p) = 0.8$. Then if it *does* rain, the accuracy score of my current credence is $s_1(0.8) = 0.04$. And if it *doesn't* rain, the accuracy score of my current credence is $s_0(0.8) = 0.64$. This scoring function, known as the Brier Score, was used in the 1950s to evaluate, and to incentivize, weather forecasters.[1]

Notice that in this case I get a higher score when my credence in the truth is lower. Other choices of f^p have the opposite orientation. All that matters is that f^p always moves in the same direction as your credence in the truth about p increases: that is, f^p is strictly monotonic. For present purposes I use the following scoring rule:

$$s_1(x) = \ln x$$

$$s_0(x) = \ln(1-x).$$

On this measure of accuracy, it is obvious that to aim at accuracy is to aim at maximizing f^p.[2]

[1] Brier 1950; Roulston 2007.

[2] Why not something simpler, like $s_1(x) = x$ and $s_0(x) = 1 - x$? Well, scoring rules ought to be *strictly proper*: any credence function uniquely maximizes its own expected score; no credence function Cr considers any other credence function Cr' to be as accurate or more accurate in expectation, where expected accuracy is calculated by the lights of Cr. This implies that for any

Table B1 Posterior $Cr(y)$ in TGL

		y	$(n, 1)$	$(n, 2)$	$(n, 3)$
o_1		0.625	0.625	0	0
o_2		0.625	0	0.625	0
o_3		0.625	0	0	0.625
r	r_1	0.667	0.667	0	0
	r_2	0.667	0	0.667	0
	r_3	0.667	0	0	0.667

Let us apply this to *The Great Lorenzo*. Let p be the proposition y that Lorenzo is lacto-pathic. And suppose that when you are choosing between experimental procedures, the value of any outcome is the accuracy, in that outcome, of your credence that Lorenzo is lacto-pathic; that is: $f^p(Cr)$. To calculate this we need your credence that Lorenzo is lacto-pathic in each outcome. We calculate this using the conditional probabilities in Table 4.4. The results are as in Table B1.

Each entry in the table indicates your confidence in the situation where the row and column heading both hold that Lorenzo is lacto-pathic. For instance, the top-left entry says that if Lorenzo is lacto-pathic and you directly put the milk first into teacup no. 1, then you end up with confidence 0.625 in the true proposition that he is lacto-pathic. Similarly, the entry corresponding to r_1 and $(n, 3)$ says that if in fact you choose to randomize, the robot puts the milk into teacup no. 1, and Lorenzo is not lacto-pathic and chooses teacup no. 3, then you will be *certain* that he is not lacto-pathic.[3] Note in particular that the entries corresponding to each r_i and y, and to each r_i and n_i, differ from those

$z \in [0, 1]$ the function $g_z(x) = zs_1(x) + (1 - z)s_0(x)$ achieves a maximum (or minimum) value over $x \in [0, 1]$ when and only when $x = z$. The functions $s_1(x) = \ln x$, $s_0(x) = \ln(1 - x)$ have this property, but the functions $s_1(x) = x$, $s_0(x) = 1 - x$ do not. Strict propriety of an accuracy measure implies other attractive properties: see e.g. Gneiting and Raftery 2007. By choosing an experiment that maximizes the expectation of the logarithmic score we are also maximizing the average information that the experiment provides in the sense of Lindley 1956.

[3] To illustrate how I calculated these values, consider the top-left entry i.e. the case corresponding to o_1 and y. What we observe in that case is the proposition p that Lorenzo chooses the correct teacup. We know from Table 4.4 that $Cr(p|y \cap o_1) = 1$ and that $Cr(p|n \cap o_1) = \frac{0.24}{0.24+0.18+0.18} = 0.4$. We also know that $Cr(y|o_1) = Cr(y) = 0.4$ because your choice of experimental procedure by itself does not affect your confidence that Lorenzo is lacto-pathic. Therefore also $Cr(n|o_1) = Cr(n) = 0.6$. The following is provable from (2.1)–(2.4):

$$Cr(y|p \cap o_1) = \frac{Cr(p|y \cap o_1)Cr(y|o_1)}{Cr(p|y \cap o_1)Cr(y|o_1) + Cr(p|n \cap o_1)Cr(n|o_1)}.$$

Running the figures through this formula gives $Cr(y|p \cap o_1) = \frac{0.4}{0.4+0.24} = 0.625$. On the Bayesian assumption (2.4), this figure represents your confidence that Lorenzo is lacto-pathic in the

Table B2 Posterior accuracy in TGL

		y	$(n,1)$	$(n,2)$	$(n,3)$
o_1		-0.470	-0.981	0	0
o_2		-0.470	0	-0.981	0
o_3		-0.470	0	0	-0.981
r	r_1	-0.405	-1.099	0	0
	r_2	-0.405	0	-1.099	0
	r_3	-0.405	0	0	-1.099

corresponding to each o_i and y, and to each o_i and n_i respectively: this is because learning that Lorenzo has passed the *randomized* version of the test makes you more confident that he is lacto-pathic than does learning that he has passed the non-randomized version of it.

We now calculate accuracy scores in each outcome.

Strictly speaking, the numbers in Table B2 could not be the news values of the corresponding outcomes: according to the definition (2.5). news value must always take a percentage value between 0 and 100. But by imposing a suitable affine transformation (e.g. $x \mapsto \frac{100x}{1.099} + 100$) we could restrict them to an appropriate range. We may then apply (2.7) and (2.9) to Table B2, using the conditional probabilities in Table 4.4 to calculate news values for o_1, o_2, o_3 and r but using your *unconditional* credences in the (n, i) i.e. using $Cr((n, i)) = 0.2$ to calculate their causal utilities.

As expected, the news value of r exceeds that of each of the o_i: applying the transformation mentioned we get $V(o_i) = 61.5$, $V(r) = 65.3$. But the causal utility of r *also* exceeds that of each of the o_i: $U(o_i) = 65.0$, $U(r) = 65.3$. Nor is this an accident. Given a range of natural ways to measure accuracy, even CDT recommends randomization to a Bayesian enquirer who cares only about maximizing accuracy.[4]

situation where he *is* lacto-pathic and so passes the test based on direct allocation of the milk to teacup 1. By a similar argument we can calculate the other entries in the table.

[4] To see this, write q for $Cr(y|p \cap o_1)$ (in the notation of n. 3), and without loss of generality consider a scoring function which is positively oriented (higher = more accurate) and where $s_1(1) = s_0(0) = 0$. Then $U(o_1) = 0.4s_1(q) + 0.2s_0(q) + 0.4s_0(0)$ and $U(r) = 0.4s_1(0.667) + 0.2s_0(0.667) + 0.4s_0(0)$. So $U(o_1) \geq U(r)$ iff $0.4s_1(q) + 0.2s_0(q) \geq 0.4s_1(0.667) + 0.2s_0(0.333)$ i.e. (dividing both sides by 0.6) iff $0.667s_1(q) + 0.333s_0(q) \geq 0.667s_1(0.667) + 0.333s_0(0.667)$. But this can never happen if the scoring rule is strictly proper (n. 2), because in that case $xs_1(x) + (1-x)s_0(x) > xs_1(z) + (1-x)s_0(z)$ if $x, z \in [0, 1]$ and $x \neq z$. Substituting $x = 0.667$ and $y = q$ then implies that $U(r) > U(o_1)$ if you respect a strictly proper scoring rule, since the fact that there is a correlation between your choice and Lorenzo's implies that $q < 0.667$. Even CDT recommends that you randomize in any of these circumstances.

So the argument at Section 4.2 does not apply if your values reflect the accuracy of your credences. Does this mean that in that case EDT and CDT agree on the value of randomization? No, because now CDT is *too* enthusiastic about randomization.

To see the idea behind this, suppose that there is a *perfect anti*-correlation between your choice of teacup and Lorenzo's guess, if he is a fraud. That is, if you choose a teacup directly, then your choice is certain to have been influenced by a subliminal cue that will *certainly* prompt Lorenzo to choose *another* teacup, if he is a fraud. But this correlation is broken if you randomize.

In this case, it is intuitively sensible *not* to randomize. By choosing directly, you are *certain* to find out that Lorenzo is a fraud, if he is. And if he is *not* a fraud, then you are certain to find that out too, because in that case he is certain to choose correctly. So choosing directly is *guaranteed* to make your post-experimental credences completely accurate. And yet absurdly, CDT *still* prefers randomization. Intuitively, this is because it ignores the non-causal anti-correlation between your direct choice and Lorenzo's guess. Since your direct choice cannot influence Lorenzo's guess, CDT cares about the possibility that Lorenzo is fraudulent but guesses correctly, even though this outcome is certain not to happen.

Let us go through a more realistic version of this case in more detail.

> *TGL2*: Like *The Great Lorenzo*, except this time you suspect that any subconscious influences on your direct choices work in the opposite direction on Lorenzo's choice, if he is a fraud. So if you directly put the milk first into teacup no. 1, say because it is very slightly brighter than the other two, Lorenzo is more likely to guess (if he does guess) no. 2 or no. 3.

We can set up Table B3 for conditional probabilities analogous to Table 4.4. We then use this table to derive the posterior probability, in each outcome, that Lorenzo is lacto-pathic, as in Table B4. Then we calculate the accuracy of your posterior credences. Table B5 records not these scores but the result of subjecting them to an affine transformation that normalizes their values to between 0 and 100.

We now calculate the news values and causal utilities of the four options on the assumption that you care about accuracy. For news values we use Tables B3 and B5 directly to give

$$V(o_1) = V(o_2) = V(o_3) = 84.5$$

$$V(r) = 76.3$$

As for causal utilities, recall that $Cr(y) = 0.4$ and $Cr((n,i)) = 0.2$ for each i; also that which cup Lorenzo chooses is (as before) causally independent of your direct choice. Combining these facts with Table B5 and (2.9):

Table B3 TGL2 probabilities of outcomes

		y	$(n,1)$	$(n,2)$	$(n,3)$
o_1		0.4	0.1	0.25	0.25
o_2		0.4	0.25	0.1	0.25
o_3		0.4	0.25	0.25	0.1
r	r_1	0.4	0.2	0.2	0.2
	r_2	0.4	0.2	0.2	0.2
	r_3	0.4	0.2	0.2	0.2

Table B4 Posterior $Cr(y)$ in TGL2

		y	$(n,1)$	$(n,2)$	$(n,3)$
o_1		0.8	0.8	0	0
o_2		0.8	0	0.8	0
o_3		0.8	0	0	0.8
r	r_1	0.667	0.667	0	0
	r_2	0.667	0	0.667	0
	r_3	0.667	0	0	0.667

$$U(o_1) = U(o_2) = U(o_3) = 74.5$$

$$U(r) = 76.3$$

So this time it is the other way around: CDT prefers to randomize, whereas EDT prefers not. As before, the point holds for a wide range of natural ways to measure probabilistic accuracy.[5]

But surely EDT has got things right. In the example as stated, the negative correlation between your choice and Lorenzo's means that if you choose a cup directly, you are more likely to catch him out if (but only if) he is a fraud. What is wrong with exploiting this? Conversely, what could be *right* about throwing away this advantage just because it isn't causal, or rather not causal in the right way (i.e. not reflective of any influence of your choice upon his)? Experiments which *do* exploit it tend to create more accurate posterior credences: as V reflects, the expected normalized accuracy of your beliefs on this matter following a direct choice exceeds 84, whereas if you use the robot, it is less than 77.

[5] By the argument at n. 4, this disagreement can be made to arise whenever the agent cares only about the accuracy of her posteriors according to a strictly proper scoring function.

Table B5 Normalized posterior accuracy in TGL2

		y	$(n,1)$	$(n,2)$	$(n,3)$
o_1		86.14	0	100	100
o_2		86.14	100	0	100
o_3		86.14	100	100	0
r	r_1	74.81	31.74	100	100
	r_2	74.81	100	31.74	100
	r_3	74.81	100	100	31.74

Of course, you could insist that there is something special about the metaphysical connection that *causal* influence sets up; that what matters about your choice of cup is only its *effect* on, only what it *does* to, the accuracy of your posterior beliefs about The Great Lorenzo. But if you do insist on that, then you inevitably pay a price in accuracy; but in the context of scientific inquiry, accuracy was meant to be the only thing that mattered.

Appendix C
Bolker–Jeffrey Representation Theorem

This book describes the content and applications of EDT as a normative theory of decision; but I mentioned in the introduction that historically EDT arose from Bolker's representation theorem. This appendix does not attempt to prove these: for formal proof the reader is better off with the original papers (Bolker 1966, 1967) or working through Jeffrey's exposition (Jeffrey 1983). Rather it attempts to lay out some intuition behind Bolker's ideas.

Suppose a robot runs the following programme: if you feed it the names of any two regions of a large plane surface, it will return at least one of them: the preferred region. What Bolker showed was analogous to this: first, that if the robot's answers to these questions satisfied certain rules, then it is as if the robot (i) has assigned an area to each region; (ii) has assigned an amount of gold to each region; (iii) when asked to choose between any two regions, always prefers that with the highest concentration of gold (i.e. the most gold per unit area). And second, that a specific range of such assignments can model the robot's behaviour in this way.

In this analogy, the plane corresponds to the set of all possible worlds, the regions to propositions, their areas to your confidence in them, and the concentration of gold in any region to its desirability or news value. The robot corresponds to you, and what Bolker showed more literally was: first, that if your preferences between propositions meet certain conditions then it is as if you have psychological states corresponding to (i) confidence in each proposition and (ii) the product of confidence and news value, such that (iii) in any choice between propositions you always prefer that with the greatest news value. And second, that if one specification of psychological states does this trick, then so does any within a specifiable range.

As I said at Section 1.2, representation theorems like this are contributions to the philosophy of mind. A representation theorem for a descriptively adequate behaviouristic theory *B* may *vindicate* whatever mentalistic notions are involved in the psychological theory *P* with which the theorem connects it, since it shows that the behaviour can always be adequately modelled by the psychology.

But what also matters is the *range* of values of the psychological parameters that model the behaviour. One might argue from a behaviouristic standpoint that there are no psychological facts about you beyond those common to all specifications of those parameters that adequately simulate your behavioural dispositions. If, for example, my behavioural dispositions are limited to bets on horses, and if they are all consequences of P given *either* (a) that I like money and am well-informed about every race, or (b) that I dislike money and am poorly informed about every race, then on this view there is no fact as to which of (a) and (b) is true.

In this exposition of the Bolker–Jeffrey Representation Theorem I'll use mainly spatial analogies to give an intuitive idea of the theory. But I'll use other analogies where appropriate – with concentration (of a solute) in connection with the averaging and impartiality axiom, and with a speedometer in connection with the uniqueness result.

C1 The Psychological Theory

Imagine a plane surface divided up into regions p, q etc.

Some regions may overlap (have points in common). The overlap of two regions p and q is itself a region, labelled $p \cap q$, the **intersection** of p and q. (For instance, there is a region consisting of the overlap of the territory of Egypt and the Sahara Desert).

Any regions p and q, overlapping or not, together form a region labelled $p \cup q$, the **union** of p and q. (For instance, the territories of Ulster and the Republic of Ireland together form the island of Ireland.) In fact, any regions $p_1, p_2 \ldots$, finite or infinite in number, form a region, which we label $\cup_n p_n$ or the union of the p_n. (For instance, the territory of the United States is the union of all its States and Territories (plus DC), including some like Alaska and Hawaii that do not adjoin any others.)

Some regions are wholly contained within others: if every point in p is also in q, then p is a **subset** of q: $p \subseteq q$. Every region is a subset of itself: so $p \subseteq p$ for any p; and it is a **proper subset** of any other region q of which it is a subset, written $p \subset q$. (For instance, the territory of Lancashire is both a subset and a proper subset of the territory of the UK.)

For every region p there is a region comprising the entire plane except p; this is the **complement** or negation of p, written $\overline{p}$. The entire plane itself is a region, labelled T or the **tautology**; its complement is the empty region, labelled F or $\emptyset$.

We can always divide the plane into regions that do not overlap but which cover it completely. Any collection Π of such regions is a **partition**. For instance, if b is the black region of a chessboard and w is the white region, then $\Pi = \{b, w\}$ is a partition of the board: all points are either black or white; no point is both. Intuitively it is clear that any region can be divided, without remainder or overlap, into exactly those regions consisting of its overlap with each member of a partition. For instance, let p comprise the four squares in the top left-hand corner of the chessboard. Then p comprises the white region that belongs to it and the black region that belongs to it, in symbols: $p = (p \cap b) \cup (p \cap w)$. Generally for any partition Π and region p we have:

$$p = \cup_{q \in \Pi} p \cap q. \tag{C1}$$

In this thought experiment each region has two further properties: *area* and *net gold*.

C1.1 The First Measure: Area

The **area** of any region p of this plane, here written $Cr(p)$, is measured by the proportion of the whole plane T that it occupies. Clearly every region p has positive or zero area and the area of T is 1. Clearly also the area of any region comprising two regions that don't overlap – whose overlap is empty – is the sum of their areas. In symbols, for any regions p and q:

$$Cr(p) \geq 0 \tag{C2}$$

$$Cr(T) = 1 \tag{C3}$$

$$p \cap q = \varnothing \rightarrow Cr(p \cup q) = Cr(p) + Cr(q). \tag{C4}$$

For any two regions p and q, we can also measure the *proportion of the area of* p that is occupied by q. Since the region of p that is occupied by q is just $p \cap q$, its area is $Cr(p \cap q)$. The area of p is obviously $Cr(p)$. Therefore if $Cr(p)$ is positive, then the proportion of the area of p that is occupied by q is the ratio of these quantities, that is, $Cr(p \cap q)/Cr(p)$. We abbreviate this with the symbol $Cr(q|p)$:

$$Cr(q|p) =_{\text{def}} Cr(p \cap q)/Cr(p). \tag{C5}$$

For instance, if region p occupies half of the plane, and region q occupies half of region p and all the rest of the plane, then $Cr(p \cap q) = .25$ and $Cr(p) = 0.5$, so $Cr(q|p) = 0.5$.

C1.2 The Second Measure: Net Gold

God, who created this world, also created equal quantities of *gold* and *anti-gold* which he distributed by burying it across the plane. Gold is valuable. Anti-gold is noxious and corrosive: 1 kg of anti-gold costs exactly as much to store or to dispose of as 1 kg of gold can fetch on the open market.

The *net gold* of a region p, written $G(p)$, is the mass in some unit of gold net of anti-gold buried under it. Since the total amounts of gold and anti-gold are equal, the net gold of the entire plane is zero: $G(T) = 0$. Clearly also, the net gold of any area comprising two regions that don't overlap is the sum of the net gold of those regions. In symbols:

$$G(T) = 0 \tag{C6}$$

$$p \cap q = \emptyset \rightarrow G(p \cup q) = G(p) + G(q). \tag{C7}$$

We know from (C1) that any region p is just the sum of its overlaps with each member of any partition Π: $p = U_{q \in \Pi}\, p \cap q$. Since net gold is additive, that is, satisfies (C7), the net gold in a region is the sum of the net gold in the overlap of that region with each member of a partition. Putting this in symbols, for any region p and any partition Π we have:

$$G(p) = \sum\nolimits_{q \in \Pi} G(p \cap q). \tag{C8}$$

The story so far. God has buried gold and anti-gold under the regions of a plane. The area of a region is the sum of the areas of its non-overlapping parts. The net gold buried under a region is the sum of the net gold buried under its non-overlapping parts.

C1.3 The Ratio of These Measures

Suppose God places you at some point in the plane. He doesn't tell you where, but he does reveal some region to which that point belongs. All that you care about is the net gold at the point where you are placed – the more the better.[1] Therefore, how pleased you would be to learn that your point lies in region p should depend on the *density* of net gold in that region (i.e. the *ratio* of its net gold to its area). If p is a region then we call the density of net gold in a region its value, written $V(p)$. So:

$$G(p) = V(p)Cr(p). \tag{C9}$$

That is, the net gold associated with any region is just the product of its area with its density of net gold. Now since (C9) holds of any region of positive area, we can substitute (C9) into (C8) to get

[1] It doesn't make obvious sense to talk about net gold at a *point*. A point has zero area; so how, in this model, could any gold or anti-gold be associated with it? But then in that case, it looks like all points have equal (i.e. zero) net gold. But in a more rigorous and abstract treatment, G is absolutely continuous with respect to Cr, so indeed zero area implies zero net gold, *but* there exists a non-zero function f (the Radon–Nikodym derivative of G with respect to Cr) taking *points* to real numbers such that for any area-measurable region p we have $G(p) = \int_p f \, dCr$. Then as with probability density functions for continuous distributions, we can understand the net gold associated with a point x to be $f(x)$.

$$V(p)Cr(p) = \sum_{q \in \Pi} V(p \cap q)Cr(p \cap q). \tag{C10}$$

This equation holds for any partition Π whose elements all have a positive area of overlap with p. For any such partition we divide through by $Cr(p)$ and apply (C5) to give

$$V(p) = \sum_{P \in \Pi} V(p \cap q)Cr(q|p). \tag{C11}$$

This is the analogue of the fundamental equation (2.7) relating news value to credence.

That completes this brief exposition of the psychological theory. We now turn to the more behaviouristic side (i.e. constraints on binary preference). Essentially the idea is this: if your preferences over regions satisfy certain axioms, there is some possible assignment of areas to those regions, and some possible distribution of gold and anti-gold on the plane, such that you are maximizing the density of net gold.

C2 The 'Behaviouristic' Theory

Although (C11) can be regarded as an abstract statement connecting the ratio V of two measures with one of them, the intended interpretation of Evidential Decision Theory is a doctrine about preferences. In making this connection, the classic expositions (Jeffrey 1983, Bolker 1967) start with axioms governing preferences amongst various prospects (here: regions) and show that for anyone whose preferences *do* satisfy those assumptions, it is *as if* there were quantities like area, net gold and density of net gold associated with each prospect, *as if* those quantities satisfied equations (C1)–(C11) and *as if* her preference between prospects was always for that with the higher density.

C2.1 Preference for news items

The basic relation by which we interpret the axioms is *preference over news items*: given any pair of news items, which one would you rather learn? Continuing with our analogy, I'll abuse notation as follows: p will henceforth mean not only some region of the plane, but also the proposition that the point you occupy is in that region, propositions being understood as sets of possible worlds as at Section 2.1.

I'll say that you **strictly prefer** p to q, $p \succ q$, if you would rather learn p than q. You are **indifferent** between p and q, $p \sim q$ if you don't care which you learn. And you **weakly prefer** p to q, written $p \succsim q$, if you *don't* strictly prefer q to p. On the assumption that you cannot strictly prefer anything to itself or strictly prefer each of two prospects to the other, it follows that for any prospects p and q we have $p \succsim p$ and either $p \succsim q$ or $p \succsim q$: weak preference is reflexive and complete.

There is one respect in which preference so understood is *non*-behaviouristic. We typically regard choices as between exclusive options: if one chooses between x and y, then one cannot have both. (More strongly, a person chooses at a time from a *partition*: you choose *exactly one* of the available prospects.) So if strict preference for x over y means that you always *choose* x in a straight choice between x and y, this makes it hard to understand preference between *compatible* propositions. What is the behaviouristic interpretation of a preference for the prospect that the card drawn is the ace of hearts over the prospect that it is *some heart or other*? If one deliberately chooses the ace of hearts, is one not *also* choosing some heart or other? This problem arises whenever we discuss preferences between compatible prospects, as several of Bolker's axioms do.

Of course, there is no difficulty in the idea that one prefers to *learn* a proposition to learning another that is compatible with it. We know what it means to *prefer learning* that a given card is the ace of hearts to learning only that it is some heart or other.

But then the theory cannot be applied to people, machines or anything else *if* we restrict our data about them to mere descriptions of choice behaviour: for although behaviour can perhaps reveal preference between exclusive possibilities, it cannot reveal preference between compatible ones. The axioms governing preference cannot be reduced to claims about behaviour or behavioural dispositions, although that is consistent with claiming that they do or should *guide* behaviour.

C2.2 The Bolker-Jeffrey Axioms

We can now state the axioms concerning preference that underlie EDT. These fall naturally into two classes, which I call ‘Substantive’ and ‘Structural’. The substantive axioms have some intuitive plausibility. The structural axioms are formal requirements that are not plausible in themselves. They are there to impose a fineness on the subject’s preferences that lets us pin down, not uniquely but to within a certain range, numerical values for the psychological parameters that the theory gives us the resources to introduce.

C2.2.1 Substantive Axioms

One natural way to think about preference is that it reflects a ranking. Any ranking must be transitive: if p is ranked at least as high as q, and q at least as high as r, then p is ranked at least as high as r. Hence the first axiom says that weak preference is transitive: if p is weakly preferred to q and q is weakly preferred to r, then p is weakly preferred to r. Formally, for any p, q and r:

BJ1: Ranking $p \succsim q \succsim r \rightarrow p \succsim r$

Preference may not satisfy this demand. However much sugar I think is in my tea, I'd weakly prefer to learn that I am about to get 1¢ from you in exchange for a little more sugar in my tea, if this amount is small enough to be undetectable. But I also strictly prefer *not* to learn that I am about to get 10¢ from you in exchange for ten times that much sugar in my tea (which would taste terrible).

The second axiom governs the relationship between specific news items and their disjunctions. It says that if p and q are incompatible, then your preferences put $p \cup q$ 'between' p and q. Formally, for any p, q we have:

BJ2: Averaging If $p \cap q = \emptyset$, then:

(i) $p \succ q \rightarrow p \succ (p \cup q) \succ q$;
(ii) $p \sim q \rightarrow p \sim (p \cup q) \sim q$.

This idea has some informal pull. If you aim to draw the ace of hearts, then you'd strictly prefer learning (p) that the card drawn was the ace of hearts to learning (q) that it was a heart *other* than the ace of hearts. It seems plausible that you'd also strictly prefer learning p to learning ($p \cup q$) that it was some heart or other; and you'd strictly prefer that to q. But this axiom is hard to interpret in terms of choice: no choice between *exclusive* options could reveal a preference for p over $p \cup q$.

The axiom is natural when we think about ratios of measures. If p and q are glasses of diluted orange cordial and the concentration of cordial in p is greater than the concentration of cordial in q, then mixing them together gives an intermediate concentration. Similarly, if the concentration of cordial is the same in both glasses, then mixing them gives the same concentration. So it is understandable that BJ2 should hold for a preference relation that we model as a ranking by the ratio of two measures.

The next axiom says that if you are indifferent between incompatible p and q, and if *some* third proposition r, incompatible with both and not ranked with either, is such that you are indifferent between (i) learning that p or r is true and (ii) learning that q or r is true, then for *any* proposition r' that is incompatible with both and not ranked with either, you are indifferent between (i) learning that p or r' is true and (ii) learning that q or r' is true. Formally, for any p, q:

BJ3: Impartiality If $p \cap q = \emptyset$, $p \sim q$ and r is such that:

(i) $p \cap r = q \cap r = \emptyset$
(ii) $p \succ r$ or $r \succ p$

(iii) $p \cup r \sim q \cup r$,

then every r' that satisfies (i) and (ii) also satisfies (iii).

For instance, suppose you just want to draw the ace of hearts. You are indifferent between learning (p) that you drew the two of clubs and (q) that you drew the king of spades. A third proposition (r) that you drew a red card is (i) incompatible with both p and q, and (ii) strictly preferred to both (because it leaves open that you drew the ace of hearts). (iii) you are indifferent between learning (a) that you drew a red card or the two of clubs and learning (b) that you drew a red card or the king of spades. It then seems plausible that any *other* proposition r' that satisfies (i) and (ii), for instance, the proposition that the card is a heart, also satisfies (iii).

Impartiality is also naturally interpretable in connection with ratios of measures. Let p and q be two irregularly shaped bodies of water (lakes, swimming pools, etc.) in which is some concentration of free chlorine. You can directly measure (by sampling) the concentration of free chlorine in each body of water – that is, mass of chlorine per unit volume of the water in p or q – but you cannot directly measure the volume of either body of water. Testing for concentration, you learn that the concentration of chlorine in p and q is the same in both. How can you tell whether they have the same volume?

Here is a way to do it. Find some other body of water in which the chlorine is at some different level of concentration from p and q. Then add the same amount of water r from this third body to each of p and q. Then mix thoroughly and compare the concentrations of chlorine in the two new bodies, $p \cup r$ and $q \cup r$.

Suppose you find that these two new bodies of water, $p \cup r$ and $q \cup r$, also have the same concentration of chlorine. Then the two original bodies p and q must have the same volume. Otherwise, the concentration of chlorine in the smaller body of water would be closer to that in r.

But if p and q have the same volume as well as the same concentration, then adding any *other* body of water to both will *also* result in new bodies of water that have the same concentration. So if $p \cup r$ and $q \cup r$ have the same concentration of chlorine for *some* r, then $p \cup r'$ and $q \cup r'$ have the same concentration for *any* r'.[2] If what I just said about concentration goes equally for the ratio of any additive measures, it is not surprising that any preference relation that we can model as a ranking by some such ratio will satisfy Impartiality.

[2] This claim is stronger than the analogue of Impartiality, which only says that if $p \cup r$ and $q \cup r$ have the same concentration of chlorine for some r that is more or less concentrated than p then $p \cup r$ and $q \cup r$ have the same concentration for any r that is more or less concentrated than p. But it follows from the latter together with the relevant analogue of Averaging.

Table C1 Impartiality

	Rain	No rain
Bet 1	A new sports car	Nothing
Bet 2	\$20,000	Nothing
Bet 3	A new sports car	One year in jail
Bet 4	\$20,000	One year in jail

Impartiality in the present system is roughly analogous to independence in Savage's:[3] it implies that if you are indifferent between two bets on a proposition that have different prizes but the same penalty, then you are indifferent between them whatever that penalty is. For instance, consider four bets on the proposition that it rains tomorrow, the payoffs as shown in Table C1.

Suppose you are indifferent (whether or not it rains) between \$20,000 and a new sports car and prefer both to nothing and also to one year in jail. Then let p, q, r and r' be the following propositions:

p: It rains and you get a new sports car.
q: It rains and you get \$20,000.
r: It doesn't rain, and you get nothing.
r': It doesn't rain, and you get one year in jail.

(The proposition that you take) Bet 1 is equivalent to $p \cup r$. Similarly, Bet 2 is equivalent to $q \cup r$, Bet 3 to $p \cup r'$ and Bet 4 to $q \cup r'$. The impartiality axiom then implies that if you are indifferent between Bet 1 and Bet 2, then you are also indifferent between Bet 3 and Bet 4.

C2.2.2 Structural Axioms

We now discuss two more technical and unintuitive axioms. They are not necessary for most applications, including those in the main part of this Element; but they are necessary for the representation theorem.

The first involves the following notions: Boolean algebra, entailment, supremum and infimum and atomlessness. We define these for our purposes as follows:

Boolean algebra: A set S of propositions is a Boolean algebra if:

(i) The set Ω of all possible worlds belongs to S.
(ii) The empty set belongs to S.
(iii) If p belongs to S, then $\Omega - p = \overline{p}$ belongs to S.
(iv) If p and q belong to S, then $p \cup q$ belongs to S.

[3] Savage 1972: 21–3; see also definition D1 and postulate P2 in Savage's end-papers.

We then define entailment as you might expect:

> **Entailment**: If p and q are propositions, then p entails q, written $p \vDash q$, if $p \subseteq q$.

Now if R is a set of propositions, we can define the supremum and infimum of that set as follows:

> **Supremum**: The supremum of R is that proposition $S(R)$ such that (a) $S(R)$ entails all propositions in R; (b) any proposition that entails all propositions in R entails $S(R)$.
>
> **Infimum:** The infimum of R is that proposition $I(R)$ such that (a) All propositions in R entail $I(R)$; (b) $I(R)$ entails any proposition that all propositions in R entail.

Intuitively, the supremum of a finite set of propositions is their conjunction and the infimum of a set of propositions is their disjunction. For instance, if $R = \{p, q\}$, where p is the proposition that this die landed on an even number and q the proposition that it landed on a number less than 5, then $S(R)$ is the proposition that it landed on 2 or on 4, and $I(R)$ is the proposition that it landed on 1, 2, 3, 4 or 6. But the definition also applies if R is infinite. If, for example, $R = \{p_1, p_2 \ldots\}$, where p_n is the proposition that the mass of the die in grams is less than $1 + \frac{1}{n}$, then $S(R)$ is the proposition that its mass in grams is at most 1 and $I(R)$ is the proposition that it is less than 2.

We now define:

> **Completeness**: A Boolean algebra S of propositions is complete if every set of propositions in S (that is, every subset of S) has an infimum and a supremum that are also in S.

For instance, if S contains: (p_1) The mass of this die is less than 2 g; (p_2) The mass of this die is less than 1+1/2 g; (p_3) the mass of this die is less than 1+1/3 g … – if S contains all these propositions, then it also contains the proposition that the mass of this die is no more than 1 gram.

Next, atomicity. Intuitively, an atom of a Boolean algebra is a maximally informative proposition: that is, it could be true (so is not $\varnothing$), and no other propositions in that algebra entail it (except $\varnothing$, which entails everything).

> **Atomic propositions**: If S is a Boolean algebra and p is a proposition in S other than $\varnothing$, then p is an atom of S if, for any q in S such that $q \vDash p$, either $q = p$ or $q = \varnothing$.

For instance, consider the propositions about the outcome of the last throw of this die: for instance, that it landed on 6, that it landed on an even number or that it did not land on 5. The set S of all such propositions, propositions whose truth or falsity supervenes on the outcome of the throw, is a Boolean algebra. The atoms of S are $p_1, p_2, \ldots, p_6$, where p_n says that the die landed on n. There is no proposition other than $\varnothing$ that (a) supervenes on the outcome of this throw and (b) is stronger than (i.e. entails but is not entailed by) one of these propositions.

A Boolean algebra is called **atomless** if none of the propositions in it are atoms of that algebra. In other words, for any proposition other than $\varnothing$, we can always find another proposition in the algebra that entails it, is not entailed by it and is also itself not $\varnothing$. It follows that if w is an individual possible world, then the proposition $\{w\}$, which specifies in complete detail one possible history, does not belong to any atomless Boolean algebra. Such a proposition is maximally informative – it entails the answer to every question – and the point of an atomless algebra is that none of its elements are maximally informative: for any proposition, there is always one that tells you more.

We now state the fourth axiom.

BJ4: Complete and atomless field: $\succsim$ is defined on a complete, atomless Boolean algebra.

There is a set of propositions such that for any two of them, the agent weakly prefers one to the other (and perhaps the other way around), and this set is a Boolean algebra, contains *no* atoms, and is complete.

Identifying such a set is not straightforward: here is a sketch of one. Suppose you are throwing an infinitely thin dart at a line 1m long: the dart will land some precise distance along it; that is, on some point xm along it, for x some real number between 0 and 1. Let S be any set of such real numbers. Then we can write A_S for the proposition that the dart lands at a point in S. For instance, if $S = \{0.2, 0.3\}$, then A_S is the proposition that the dart lands either exactly 20 cm or exactly 30 cm along; if S is the interval $(0.2, 0.9)$ – that is, the set of all numbers between 0.2 and 0.9 – then A_S is the proposition that the dart lands strictly between 20 cm and 90 cm along the line.

We can define a notion of size, called *measure*, such that for many such sets – and for all intervals – the measure of that set is the proportion of the length of the line that it takes up.

For instance, the measure of $\{0.2, 0.3\}$ is zero; the measure of the interval $(0.2, 0.9)$ is 0.7; and the measure of (say) $(0.2, 0.9) \cup \{0.95\}$ – the set of all points that are either between 20 cm and 90 cm or exactly 95 cm along the line – is also 0.7. It is a celebrated and surprising fact that there are sets of points that *cannot* be assigned a measure, not if the measure is going to satisfy certain intuitive properties of size.[4] But most sets of interest do have a measure, and we can form the set M of all such *measurable* subsets of the line.

[4] For an example see Capinski and Kopp 2004: 301–2.

Now we can define the **symmetric difference** of two elements m_1 and m_2 of M as follows: it is the set $\boldsymbol{m_1 \Delta m_2} =_{\text{def.}} (m_1 - m_2) \cup (m_2 - m_1)$, the set of all points in one of these sets but not the other. And we define the **indiscernibility** relation $\approx$ on such subsets as follows: $m_1 \approx m_2$ if and only if the measure of $m_1 \Delta m_2$ is zero. It is easy to see that indiscernibility is an equivalence relation. We can therefore define the quotient set $M_{\approx}$ as the set of equivalence classes of $\approx$. This is a set of subsets of M – that is, a set of sets of measurable subsets of $[0, 1]$ – which is a complete and atomless Boolean algebra. In effect, it consists of what you get from the set of measurable subsets of $[0, 1]$ when you identify indiscernible sets. So we are identifying, for example, $(0.2, 0.9) \cup \{0.95\}$ with $(0.2, 0.9)$; and we are identifying all countable sets, for example, $\{0.2, 0.3\}$, $\{0.95\}$ and $\{1/n\}_{n=1}^{\infty}$, with one another and also with the empty set.

Is there an intuitive way to think about this construction? Imagine that you're looking at the line, but your vision is imperfect: you can only see regions that have positive measure, not, for example, particular points or countable sets of points. Moreover, you cannot distinguish indiscernible regions. Your sight is 'fuzzy': you cannot see any particular region, but you can be described as seeing classes of such regions. What you 'see' are the equivalence classes of the indiscernibility relation. These sets then form a complete atomless Boolean algebra over which preference may be defined consistently with BJ4.[5]

Finally, Continuity. Pick a point S on the surface of the earth. The following may be true: for any range of temperatures around that of S, we can always find a point T such that all the points that are closer to S than T have a temperature within that range.

For instance, suppose the current temperature at S is exactly 54.5 degrees Celsius. Then there is a number X_1 such that all the points within X_1 metres of S have temperature strictly between 54.4 and 54.6 degrees. There is a smaller number X_2 such that all points within X_2 metres of S have a temperature strictly between 54.49 and 54.51 degrees Celsius. There is a still smaller number X_3 such that all points within X_3 metres of S have a temperature between 54.499 and 54.501 degrees Celsius … If this condition holds for arbitrarily close approximations to S's temperature, then the relation *x is at least as hot as y* is *continuous* at point S relative to distance from S.

The continuity axiom says that this holds for preference with respect to entailment. For any proposition p lying between two others q and r in the

[5] This model implies much greater visual acuity than anyone has, because visual indistinguishability isn't really transitive: there is some positive Δ and integer $N > 0$ such that you cannot distinguish, for example, $(0, a)$ from $(0, a + \Delta)$ for any $a < 1 - N\Delta$, but also such that you can distinguish $(0, a)$ from $(0, a + N\Delta)$. Indiscernibility (as I'm calling it) is transitive in the present context because countable unions of sets of zero measure also have zero measure.

preference ranking, any proposition that is sufficiently 'close' to p in terms of entailment *also* lies between q and r in that ranking.

To make this formal, we need the notion of a chain.

> **Chain**: A chain of propositions is a set C of propositions such that for any $c_1, c_2 \in C$ either $c_1 \vDash c_2$ or $c_2 \vDash c_1$.

An example would be any set $\{p, p \cap q\}$. An example of an infinite chain is the set $R = \{p_1, p_2 \ldots\}$, where p_n is the proposition that the mass of some object in grams is less than $1 + \frac{1}{n}$. In terms of our model involving gold distributed over a surface, a chain would be a collection of nested regions (i.e. such that for any two regions, one is contained in the other).

The continuity axiom is as follows:

> **BJ5: Continuity** Suppose $a \succ S(C) \succ b$, where $S(C)$ is the supremum of a chain C. Then some $d \in C$ is such that for any $e \in C$, if $e \vDash d$, then $a \succ e \succ b$. Similarly suppose $a \succ I(C) \succ b$, where $I(C)$ is the infimum of a chain C. Then some $d \in C$ is such that for any $e \in C$, if $d \vDash e$, then $a \succ e \succ b$.

The analogy in terms of net gold: if we have a nested set of regions, then by moving to smaller and smaller regions, we can find a region such that any smaller region in the set has a density of net gold that is as close as you like to the density of the innermost region.

Those are the axioms for preference. For convenience I state them all here:

> **BJ1: Ranking** $p \succsim q \succsim r \rightarrow p \succsim r$
> **BJ2: Averaging** If $p \cap q = \varnothing$, then:
>
> (i) $p \succ q \rightarrow p \succ (p \cup q) \succ q$
> (ii) $p \sim q \rightarrow p \sim (p \cup q) \sim q$.
>
> **BJ3: Impartiality** If $p \cap q = \varnothing, p \sim q$ and r is such that:
>
> (i) $p \cap r = q \cap r = \varnothing$
> (ii) $p \succ r$ or $r \succ p$
> (iii) $p \cup r \sim q \cup r$,
> then every r' that satisfies (i) and (ii) also satisfies (iii).
>
> **BJ4: Complete and atomless field** The weak preference relation $\succsim$ is defined on a complete, atomless Boolean algebra.
>
> **BJ5: Continuity** Suppose that $a \succ S(C) \succ b$, where $S(C)$ is the supremum of a chain C. Then some $d \in C$ is such that for any $e \in C$, if $e \vDash d$,

then $a \succ e \succ b$. Similarly suppose that $a \succ I(C) \succ b$, where $I(C)$ is the infimum of a chain C. Then some $d \in C$ is such that for any $e \in C$, if $d \vDash e$, then $a \succ e \succ b$.

These five axioms characterize the preference relation.

C3 The Representation Theorem

The representation theorem is in two parts. The existence part says that if these axioms hold, then we can assign two numbers G and Cr to each proposition in such a way that a simple formula involving these numbers determines, for any two propositions, which is weakly preferred. The uniqueness part says that there is more than one way to make this assignment; but all assignments relate to one another in a certain way.

C3.1 *Existence*[6]

Suppose $\succsim$, $\succ$ and $\sim$ defined on a complete atomless field S of propositions from which $\varnothing$ is removed, and that they satisfy BJ1–5. Then there exist functions Cr and G, each assigning a real number to each proposition in S such that for any $p, q \in S$

(i) $Cr(p) > 0$
(ii) $Cr(\Omega) = 1$
(iii) If $p \cap q = \varnothing$, then $Cr(p) + Cr(q) = Cr(p \cup q)$
(iv) $G(\Omega) = 0$
(v) If $p \cap q = \varnothing$, then $G(p) + G(q) = G(p \cup q)$
(vi) $p \succsim q$ if and only if $\frac{G(p)}{Cr(p)} \geq \frac{G(q)}{Cr(q)}$

In this case the pair (G, Cr) *represents* the ranking $\succsim$.

The existence theorem has an abstract significance independently of the interpretation of Cr and G; but its philosophical importance is that there *are* natural psychological interpretations of Cr (as subjective probability) and of G/Cr (as news value).

C3.2 *Uniqueness*[7]

Informally, we can think of the uniqueness result as answering the following question. Suppose some assignment of Cr and G to propositions in our algebra induces a ranking of those propositions by density (i.e. by G/Cr). Then what

---

[6] This is Bolker's Theorem 6.12 (1966: 310).

[7] This is Bolker's Theorem 3.6 (1966: 302). NB: in the August 1966 *Transactions of the American Mathematical Society*, pp. 301 and 302 are printed the wrong way around.

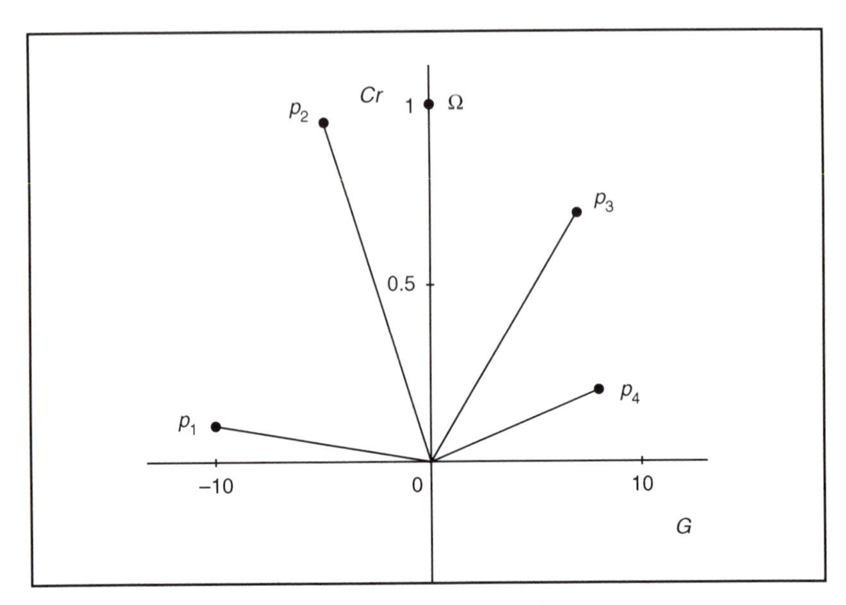

Figure C1 News value speedometer

other assignments Cr' and G' consistent with (i)–(vi) in the existence theorem are such that G'/Cr' induces the same ranking?

We can get some intuition about this by looking through the lens of elementary linear algebra.

In Figure C1, points are propositions, and their news values are measured as if on a speedometer: the news value of a proposition p, that is, the ratio $G(p)/Cr(p)$, is higher as one moves clockwise. So any preference relation that this diagram models must satisfy $p_4 \succ p_3 \succ \Omega \succ p_2 \succ p_1$. Notice that $Cr(\Omega) = 1$ and that no proposition lies above $Cr = 1$ or below $Cr = 0$.

Now the question about uniqueness becomes: what transformations of the plane in Figure C1 preserve (i)–(vi)? That is, suppose (G, Cr) represents $\succsim$. For any transformation T, if we take all propositions in the algebra, and for each such proposition p apply the transformation T to the vector $(G(p), Cr(p))$, we get a new vector $(G^T(p), Cr^T(p))$. So the question is: what T are such that (G^T, Cr^T) also represents $\succsim$?

Inspection of (i)–(vi) and visual intuition suggest that T satisfies:

$$Cr^T(p) > 0 \text{ for any } p \text{ because of (i)} \tag{C12}$$

$$Cr^T(\Omega) = 1 \text{ because of (ii)} \tag{C13}$$

$$G^T(\Omega) = 0 \text{ because of (iv)} \tag{C14}$$

$$T \text{ preserves orientation about the origin because of (vi).} \tag{C15}$$

Visual intuition further suggests that if 'enough' propositions are in the algebra, then any transformations satisfying (C15) are representable by 2×2 matrices with positive determinant (i.e. there must be real numbers a, b, c, d) such that:

$$\begin{pmatrix} G^T \\ Cr^T \end{pmatrix} = \begin{pmatrix} a & b \\ c & d \end{pmatrix} \begin{pmatrix} G \\ Cr \end{pmatrix}, \; ad - bc > 0 \tag{C16}$$

It then follows by (C13), (C14) and (C16) that $b = 0, d = 1$ and hence from (C16) again that $a > 0$. Then, by (C12) and (C16):

$$0 < cG(p) + Cr(p) \text{ for any } p \tag{C17}$$

Since $\frac{G(p)}{Cr(p)} = V(p)$:

$$-\frac{1}{\sup_p V(p)} \leq c \leq -\frac{1}{\inf_p V(p)}^{8} \tag{C18}$$

Since $Cr^T(p) = cG(p) + Cr(p)$ for any p, (C18) means that the preference ranking determines credence uniquely if c can only take the value zero (i.e. if V is unbounded above and below).

Putting all this together and expressing it in terms of credences and news values, we find that if $(G, Cr) = (VCr, Cr)$ represents an agent's preference, then if any other news value and credence functions V', Cr' are such that $(V'Cr', Cr')$ also represents them, then there must be $a > 0$, $c \in \left[-\frac{1}{\sup_p V(p)}, -\frac{1}{\inf_p V(p)}\right]$ such that:

$$V' = \frac{aV}{1 + cV} \tag{C19}$$

$$Cr' = (1 + cV)Cr. \tag{C20}$$

Of course, all I give here is an intuition for this claim. But Bolker proved it and its converse. He proved that the psychological representation of preference established by the existence part of the theorem is unique up to the transformations (C19) and (C20).

[8] Proof: it follows from $c > -\frac{Cr(p)}{G(p)} = -\frac{1}{V(p)}$ if $G(p) > 0$ and $c < -\frac{Cr(p)}{G(p)} =$ if $G(p) < 0$. Since these constraints hold for any p the most binding inequalities they imply are those at (C18).

References

Ahmed, A. 2014a. *Evidence, Decision and Causality*. Cambridge University Press.

2014b. Dicing with death. *Analysis* 74 (4): 587–92.

2016. Review of Lara Buchak, *Risk and Rationality. British Journal for the Philosophy of Science Review of Books*. Available online at: https://bjpsbooks.wordpress.com/2016/08/22/lara-buchak-risk-and-rationality.

2018. Self-control and hyperbolic discounting. In J.-L. Bermúdez (ed.), *Self-Control, Decision Theory and Rationality*. Cambridge University Press: 96–120.

2020. Equal opportunities in Newcomb's Problem and elsewhere. *Mind* 129 (515): 867–86.

and H. Price. 2012. Arntzenius on 'Why ain'cha rich?' *Erkenntnis* 77 (1): 15–30.

and J. Spencer. 2020. Objective value is always Newcombizable. *Mind* 129 (516): 1157–92.

Ainslie, G. 1991. Derivation of 'rational' economic behaviour from hyperbolic discount curves. *American Economic Review* 81 (2): 334–40.

Allais, M. 1953. Le comportement de l'homme rationnel devant le risque: critique des postulats et axiomes de l'école Américaine. *Econometrica* 21 (4): 503–46.

Bales, A. (unpublished). Decision and Dependence: A Defence of Causal Decision Theory. Doctoral dissertation, University of Cambridge.

2018. Richness and rationality: Causal Decision Theory and the WAR argument. *Synthese* 195 (1): 259–67.

Bénabou, R. and J. Tirole. 2011. Identity, morals and taboos: beliefs as assets. *Quarterly Journal of Economics* 126 (2): 805–55.

Bermúdez, J. L. 2018. Does Newcomb's Problem actually exist? In A. Ahmed (ed.), *Newcomb's Problem*. Cambridge University Press: 19–41.

Bernoulli, D. 1738. *Specimen Theoriae Novae de Mensura Sortis. Commentarii Academiae Scientiarum Imperialis Petropolitanae* V: 175–92. Reprinted as 'Exposition of a new theory on the measurement of risk', tr. L. Sommer. *Econometrica* 22 (1) (1954): 23–36.

Bodner, R. and D. Prelec. 2003. Self-signalling and diagnostic utility in everyday decision making. In I. Brocas and J. D. Carrillo (eds), *The Psychology of Economic Decisions, vol. I: Rationality and Well-Being*. Oxford University Press: 105–23.

Bolker, E. 1966. Functions resembling quotients of measures. *Transactions of the American Mathematical Society* 124 (2): 292–312.

1967. A simultaneous axiomatization of utility and subjective probability. *Philosophy of Science* 34 (4): 333–40.

Bradley, R. 2017. *Decision Theory with a Human Face*. Cambridge University Press.

and H. O. Stefánsson 2016. Desire, expectation, and invariance. *Mind* 125 (499): 691–725.

Brier, G. W. 1950. Verification of forecasts expressed in terms of probabilities. *Monthly Weather Review* 78 (1): 1–3.

Broome, J. 1991. *Weighing Goods*. Oxford: Blackwell.

Buchak, L. 2013. *Risk and Rationality*. Oxford University Press.

Capinski, M. and E. Copp. 2004. *Measure, Integral and Probability*. 2nd ed. London: Springer.

Christensen, D. 1992. Confirmational holism and Bayesian epistemology. *Philosophy of Science* 59 (4): 540–57.

Chu, F. and J. Y. Halpern. 2004. Great expectations. Part II: generalized expected utility as a universal decision rule. *Artificial Intelligence* 159 (1–2): 207–29.

De Finetti, B. 1937. La prévision: ses lois logiques, ses sources subjectives. *Annales de l'Institut Henri Poincaré* 7: 1–68.

Diamond, P. 1967. Cardinal welfare, individualistic ethics and interpersonal comparison of utility: comment. *Journal of Political Economy* 75 (5): 765–6.

Dove, M. R. 1993. Uncertainty, humility and adaptation in the tropical forest: the agricultural augury of the Kantu'. *Ethnology* 32 (2): 145–67.

Dummett, M. A. E. 1964. Bringing about the past. *Philosophical Review* 73 (3): 338–59.

Eells, E. 1982. *Rational Decision and Causality*. Cambridge University Press.

Elga, A. 2020. Newcomb University: A play in one act. *Analysis* 80 (2): 212–21.

Ellsberg, D. 1961. Risk, ambiguity and the Savage axioms. *Quarterly Journal of Economics* 75 (4): 643–69.

Fernandes, A. 2017. A deliberative approach to causation. *Philosophy and Phenomenological Research* 95 (3): 686–708.

Fisher, R. A. 1935. *The Design of Experiments*. Edinburgh: Oliver and Boyd.

Flew, A. G. N. 1954. Can an effect precede its cause? *Proceedings of the Aristotelian Society Supplementary Volume* 28: 47–62.

Gallow, D. 2021. Riches and rationality. *Australasian Journal of Philosophy* 99 (1): 114-129.

Garber, D. 1983. Old evidence and logical omniscience in Bayesian confirmation theory. In J. Earman (ed.), *Testing Scientific Theories* (Midwest Studies in the Philosophy of Science, vol. X). Minneapolis: University of Minnesota Press, 99–131.

Gibbard, A. and W. L. Harper. 1978. Counterfactuals and two kinds of expected utility. In C. Hooker, J. Leach, and E. McClennen (eds), *Foundations and Applications of Decision Theory*. Dordrecht: Riedel: 125–62. Reprinted in P. Gärdenfors and N.-E. Sahlin (eds), *Decision, Probability and Utility*. Cambridge University Press 1988: 341–76.

Gneiting, T. and A. E. Raftery. 2007. Strictly proper scoring rules, prediction, and estimation. *Journal of the American Statistical Association* 102 (477): 359–78.

Grafstein, R. 1991. An evidential decision theory of turnout. *American Journal of Political Science* 35 (4): 989–1010.

1999. *Choice-Free Rationality: A Positive Theory of Political Behaviour*. Ann Arbor: University of Michigan Press.

2018. Newcomb's problem is everyone's problem: making political and economic decisions when behaviour is interdependent. In A. Ahmed (ed.), *Newcomb's Problem*. Cambridge University Press: 96–114.

Greene, P. 2018. Success-first decision theories. In A. Ahmed (ed.), *Newcomb's Problem*. Cambridge University Press: 115–37.

Holton, R. J. 2016. Addiction, self-signalling and the deep self. *Mind and Language* 31 (3): 300–13.

Unpublished. Self-deception and the moral self.

Hume, D. 1738 [1949]. *Treatise of Human Nature*. London: J. M. Dent.

Huttegger, S. M. 2017. *The Probabilistic Foundations of Rational Learning*. Cambridge University Press.

Icard, T. 2021. Why be random? *Mind* 130 (517): 111-39

Irwin, T. 1985. (Tr.) *Aristotle: Nicomachean Ethics*. Indianapolis: Hackett.

Jeffrey, R. C. 1977. A note on the kinematics of preference. *Erkenntnis* 11 (1): 135–41.

1983. *The Logic of Decision*. 2nd ed. Chicago: University of Chicago Press.

Joyce, J. M. 1999. *Foundations of Causal Decision Theory*. Cambridge University Press.

2011. The development of subjective Bayesianism. In D. M. Gabbay, S. Hartmann, and J. Woods (eds), *Handbook of the History of Logic*, vol. 10. Oxford: Elsevier, 415–75.

2012. Regret and instability in Causal Decision Theory. *Synthese* 187 (1): 123–45.

2018. Deliberation and stability in Newcomb problems and pseudo-Newcomb problems. In A. Ahmed (ed.), *Newcomb's Problem*. Cambridge University Press: 138–59.

Kadane, J. and T. Seidenfeld. 1990. Randomization in a Bayesian perspective. *Journal of Statistical Planning and Inference* 25 (3): 329–45.

Kahnemann, D. and A. Tversky. 1984. Choices, values and frames. APA award address. *American Psychologist* 39 (4): 341–50.

Kreps, D. 1988. *Notes on the Theory of Choice*. Boulder, CO: Westview Press.

Levinstein, B. and N. Soares. 2020. Cheating Death in Damascus. *Journal of Philosophy* 117 (5): 237–66.

Lewis, C. S. 1943. *The Screwtape Letters*. New York: MacMillan.

Lewis, D. K. 1976. The paradoxes of time travel. *American Philosophical Quarterly* 13 (2): 145–52. Reprinted in his *Philosophical Papers*, vol. 2. Oxford University Press 1986: 67–80.

1979. Prisoners' dilemma is a Newcomb problem. *Philosophy and Public Affairs* 8 (3): 235–4.

1981a. Causal Decision Theory. *Australasian Journal of Philosophy* 59 (1): 5–30. Reprinted with a postscript in his *Philosophical Papers* vol. 2. Oxford University Press 1986: 305–39.

1981b. Why ain'cha rich? *Noûs* 15 (3): 377–80.

1986. *On the Plurality of Worlds*. Oxford University Press.

1988. Desire as belief. *Mind* 97 (387): 323–32.

1996. Desire as belief II. *Mind* 105 (418): 303–13.

Lindley, D. V. 1956. On a measure of the information provided by an experiment. *Annals of Mathematical Statistics* 27 (4): 986–1005.

List, C. 2014. Free will, determinism, and the possibility of doing otherwise. *Noûs* 48 (1): 156–78.

Loewer, B. 2007. Counterfactuals and the second law. In R. Corry and H. Price (eds), *Causation, Physics, and the Constitution of Reality*. Oxford University Press: 293–326.

Luce, R. D. and P. Suppes. 2004. Representational measurement theory. In J. Wixted and H. Pashler (eds), *Stevens' Handbook of Experimental Psychology*, vol. 4. 3rd ed. New York: Wiley, 1–41.

Meek, C. and C. Glymour. 1994. Conditioning and intervening. *British Journal for the Philosophy of Science* 45 (4): 1000–21.

Milne, P. and G. Oddie. 1991. Act and value: expectation and the representation of moral theories. *Theoria* 57 (1–2): 42–76.

Moore, O. K. 1957. Divination – a new perspective. *American Anthropologist* 59 (1): 69–74.

Moscati, I. 2019. *Measuring Utility: From the Marginal Revolution to Behavioural Economics*. Oxford University Press.

Nozick, R. 1970. Newcomb's problem and two principles of choice. In N. Rescher (ed.), *Essays in Honor of Carl G. Hempel*. Dordrecht: D. Reidel, 114–46. Reprinted in P. Moser (ed.), *Rationality in Action: Contemporary Approaches*. Cambridge University Press 1990: 207–34.

Pasnau, R. 2020. Medieval modal spaces. In *Proceedings of the Aristotelian Society Supplementary Volume* 94: 225–54.

Pearl, J. 2000. *Causality: Models, Reasoning, and Inference*. Cambridge University Press.

Forthcoming. Causal and counterfactual inference. In M. Knauff and W. Spohn (eds), *Handbook of Rationality*. Cambridge, MA: Massachusetts Institute of Technology Press.

Peterson, M. 2017. *An Introduction to Decision Theory*. 2nd ed. Cambridge University Press.

Pettigrew, R. 2019. *Choosing for Changing Selves*. Oxford University Press.

Poundstone, W. 2014. *How to Predict the Unpredictable: The Art of Outsmarting Almost Everyone*. London: Oneworld.

Price, H. 1989. Defending desire-as-belief. *Mind* 98 (389): 119–27.

1991. Agency and probabilistic causality. *British Journal for the Philosophy of Science* 42 (2): 157–76.

and Y. Liu. 2018. 'Click!' bait for causalists. In A. Ahmed (ed.), *Newcomb's Problem*. Cambridge University Press: 160–79.

Quattrone, G. A. and A. Tversky. 1986. Self-deception and the voter's illusion. In J. Elster (ed.), *The Multiple Self*. Cambridge University Press: 35–58.

Quine, W. V. O. 1981. Things and their place in theories. In his *Theories and Things*. Cambridge, MA: Harvard University Press: 1–23.

Ramsey, F. P. 1926. Truth and probability. In his *Philosophical Papers*, ed. D. H. Mellor. Cambridge University Press 1990: 52–94.

Rothfus, G. J. 2020. Dynamic consistency in the logic of decision. *Philosophical Studies* 177: 3923–34.

Roulston, M. S. 2007. Performance targets and the Brier score. *Meteorological Applications* 14 (2): 185–94.

Savage, L. J. 1972. *The Foundations of Statistics*. 2nd ed. New York: Dover.

Seaman, B. T. 2002. What if Nancy Reagan's astrologer had sued? An essay. *Nova Law Review* 27 (2): 277–88.

Sen, A. K. 1971. Choice functions and revealed preference. *Review of Economic Studies* 38 (3): 307–17.

Simpson, E. H. 1951. The interpretation of interaction in contingency tables. *Journal of the Royal Statistical Society* (Series B) 13 (2): 238–41.

Skyrms, B. 1984. *Pragmatics and Empiricism*. New Haven: Yale University Press.

Soares, N. and B. Fallenstein. 2015. Toward idealized decision theory. arXiv: 1507.01986 [cs.AI].

Soares, N. and E. Yudkowsky. 2018. Functional Decision Theory: a new theory of instrumental rationality. arXiv: 1710.05060.

Speck, F. G. 1935. *Naskapi*. Norman: University of Oklahoma Press.

Spencer, J. and I. Wells. 2019. Why take both boxes? *Philosophy and Phenomenological Research* 99 (1): 27–48.

Stalnaker, R. 1972. Letter to David Lewis. In W. L. Harper, R. Stalnaker, and G. Pearce (eds), *Ifs: Conditionals, Belief, Decision, Chance, and Time*. Dordrecht: D. Reidel, 1980: 151–2.

Swijtink, Z. G. 1982. A Bayesian argument for randomization. *PSA: Proceedings of the Biennial Meeting of the Philosophy of Science Association*, 1 (1982): 159–68.

Von Neumann, J. and R. Morgenstern. 1953. *Theory of Games and Economic Behaviour*. 3d ed. Princeton: Princeton University Press.

Wang, Z., B. Xu, and H.-J. Zhou. 2014. Social cycling and conditional responses in the rock-paper-scissors game. *Scientific Reports* 4 (article no. 5830): 1–7.

Weirich, P. 2001. *Decision Space: Multidimensional Utility Analysis*. Cambridge University Press.

Wells, I. 2019. Equal opportunity and Newcomb's Problem. *Mind* 128 (510): 429–57.

Wittgenstein, L. 1953 [2009]. *Philosophical Investigations*. 4th ed. Tr. G. E. M. Anscombe, P. M. S. Hacker, and J. Schulte. Malden: Wiley.

Worrall, J. 2007. Why there's no cause to randomize. *British Journal for the Philosophy of Science* 58 (3): 451–88.

Yudkowsky, E. 2010. *Timeless Decision Theory*. San Francisco: The Singularity Institute.

Acknowledgements

I want first to thank my family – Frisbee, Isla, Iona and Skye – for their affection, patience and understanding during the writing of this book. And thank you very much to Iona Ahmed for the pictures.

I wrote this book in 2019–20 during leave funded by the Effective Altruism Foundation (EAF). I am very grateful to the EAF for its support.

For helpful discussion of decision theory, I wish to thank Brad Armendt, Alexander Bird, Kevin Blackwell, Ethan Bolker, Melissa Fusco, William Harper, Richard Holton, Simon Huttegger, James M. Joyce, Boris Kment, Ben Levinstein, Richard Pettigrew, Gerard Rothfus, Brian Skyrms and Jack Spencer. For extensive and very helpful written comments I am especially grateful to Adam Elga, Daniel Herrmann, Calum McNamara and two anonymous referees for Cambridge University Press. I am also most grateful to Martin Peterson for his encouragement and patience throughout this process.

Hugh Mellor, who died on 21 June 2020, made lasting and important contributions to the foundations of decision theory, quite aside from his groundbreaking work on causation, chance and time. During his time as Professor, the Cambridge Philosophy Faculty owed much of its distinctive character and reputation to the brilliance of his thought, the directness and vitality of his manner and the clarity and simplicity of his writing. He was a great philosopher, a great teacher and a great friend, and I wish to dedicate this Element to his memory.

Cambridge Elements

Decision Theory and Philosophy

Martin Peterson
Texas A&M University

Martin Peterson is Professor of Philosophy and Sue and Harry E. Bovay Professor of the History and Ethics of Professional Engineering at Texas A&M University. He is the author of four books and one edited collection, as well as many articles on decision theory, ethics and philosophy of science.

About the Series

This Cambridge Elements series offers an extensive overview of decision theory in its many and varied forms. Distinguished authors provide an up-to-date summary of the results of current research in their fields and give their own take on what they believe are the most significant debates influencing research, drawing original conclusions.

Cambridge Elements

Decision Theory and Philosophy

Elements in the Series

A full series listing is available at: www.cambridge.org/EDTP

Printed in the United States
by Baker & Taylor Publisher Services